Electric Vehicles and Distributed Generation – Microgrid

RIVER PUBLISHERS SERIES IN POWER

The "River Publishers Series in Power" is a series of comprehensive academic and professional books focussing on the theory and applications behind power generation and distribution. The series features content on energy engineering, systems and development of electrical power, looking specifically at current technology and applications.

The series serves to be a reference for academics, researchers, managers, engineers, and other professionals in related matters with power generation and distribution.

Topics covered in the series include, but are not limited to:

- Power generation;
- Energy services;
- Electrical power systems;
- Photovoltaics;
- Power distribution systems;
- Energy distribution engineering;
- Smart grid;
- Transmission line development.

For a list of other books in this series, visit www.riverpublishers.com

Electric Vehicles and Distributed Generation – Microgrid

Editors

M. Nandhini Gayathri

School of EEE, SASTRA Deemed University, Thanjavur, India

Sanjeevikumar Padmanaban

Department of Electrical Engineering, University of South-Eastern Norway, Norway

River Publishers

Routledge
Taylor & Francis Group
NEW YORK AND LONDON

Published 2025 by River Publishers
River Publishers
Alsbjergvej 10, 9260 Gistrup, Denmark
www.riverpublishers.com

Distributed exclusively by Routledge
605 Third Avenue, New York, NY 10017, USA
4 Park Square, Milton Park, Abingdon, Oxon OX14 4RN

Electric Vehicles and Distributed Generation – Microgrid / by M. Nandhini Gayathri, Sanjeevikumar Padmanaban.

Routledge is an imprint of the Taylor & Francis Group, an informa business

ISBN 978-87-7004-614-5 (hardback)
ISBN 978-87-7004-805-7 (paperback)
ISBN 978-87-7004-616-9 (online)
ISBN 978-87-7004-615-2 (master ebook)

Contents

Preface xi

List of Figures xvii

List of Tables xxi

List of Contributors xxiii

List of Abbreviations xxvii

1 Energy Storage Systems for Transportation Electrification 1
Chandan Chetri, Akash Samanta, Latha Anekal, and Sheldon Williamson
1.1 Introduction to Energy Storage System for EVs 2
1.2 Widely used Energy Storage Systems 3
1.2.1 Lithium-ion battery 3
1.2.1.1 Lithium nickel manganese cobalt oxide (NMC) battery 3
1.2.1.2 Lithium nickel cobalt aluminum (NCA) oxide 4
1.2.1.3 Lithium iron phosphate (LFP) battery . . . 5
1.2.1.4 Lithium manganese oxide (LMO) battery . 5
1.2.1.5 Lithium titanium oxide (LTO) battery . . . 6
1.3 New and Potential Future Use Energy Storage Systems . . . 7
1.3.1 Fuel cell 7
1.3.2 Solid-state battery 10
1.3.3 Lithium-sulfur battery 10
1.3.4 Nickel-metal hydride (NiMH) battery 11
1.3.5 Supercapacitors 12
1.3.6 Flywheel energy storage 13
1.3.7 Thermal energy storage 15

1.3.8 Graphene-based batteries 16
1.3.9 Flow batteries 17
1.3.10 Aluminum-ion batteries 18
1.3.11 Zinc-air batteries 19
1.3.12 Sodium-ion batteries 19
1.3.13 Molten salt batteries 20
1.3.14 Compressed air energy storage 21
1.3.15 Thermal electric generators 22
1.4 Hybrid Energy Storage System (HESS) 23
1.4.1 Battery and supercapacitor HESS 23
1.4.2 Topologies of battery and superconductor HESS . . 24
1.4.2.1 Passive HESS 24
1.4.2.2 Active HESS 26
1.4.2.3 Semi-active HESS 26
1.4.2.4 Parallel semi-active HESS 27
1.4.2.5 Battery semi-active HESS 27
1.4.2.6 Capacitor semi-active HESS 28
1.4.2.7 Full-active HESS 29
1.4.2.8 Modular multilevel full-active HESS . . . 30
1.5 Conclusion 30

2 Electric Vehicle Battery Chargers **39**
Roscia Jeya Shiney Jayakumar and Allwyn Clarence Asis Arul
2.1 Introduction 40
2.2 Electric Vehicle (EV) 41
2.2.1 Battery charging modes for EVs 42
2.3 EV Charging Methods 43
2.3.1 Battery exchange (BSS) 43
2.3.2 Wireless power transfer (WPT) 44
2.3.3 Overnight depot charging 45
2.3.4 Pantograph charging 45
2.3.5 Induction charging 45
2.4 EV Converter Topology 46
2.4.1 Recent developments of DC–DC Converter 47
2.4.2 Recent developments of AC–DC converter 49
2.5 EVs Battery Chargers Architectures 50
2.5.1 Power factor correction (PFC) stage circuit topologies 51
2.5.2 Battery current regulation stage circuit topologies . . 51

2.6 Battery Charger Controllers 54
2.7 Electric Vehicle Charging Challenges and Possible Solutions 56
2.8 Smart Battery Charger 58
2.9 Vehicle to Grid (V2G) 60
2.10 Conclusion 62

3 Wireless Charging Schemes for EVs **69**
Khalid Yahya, Mohamed Salem, Adel E. M. Yahya, Nassim A. Iqteit, and N. Prabaharan
3.1 Introduction 70
3.2 Magnetic Resonance-based Wireless Charging 71
3.3 Inductive Wireless Charging 74
3.3.1 Inductive charging schemes and topologies 74
3.3.1.1 Performance analysis of inductive charging for EVs 75
3.3.2 Capacitive wireless charging for EVs 76
3.3.2.1 Performance analysis of capacitive wireless charging for EVs 78
3.3.3 Dynamic wireless charging for EVs 79
3.3.3.1 Performance analysis of dynamic wireless charging for EVs 81
3.4 Wireless Charging Roadway Systems 82
3.4.1 Performance analysis of wireless charging roadway systems for EVs 83
3.5 Comparative Analysis of Wireless Charging Schemes 83
3.6 Conclusion and Future Directions 85

4 Power Quality Assessment and Improvement of Grid-tied Distributed Generation Systems **89**
M. Yu. Artemenko, V. V. Chopyk, V. M. Mykhalskyi, and I. A. Shapoval
4.1 Introduction 90
4.2 Improvement of the Multiphase Instantaneous Power Theory 94
4.3 Development of Integral Power Theory of Multiphase Power Supply Systems 108
4.4 Assessment and Improvement of Electric Energy Quality . . 122
4.5 Conclusions 139

5 Evolution of Microgrid, Hierarchical Architecture of Microgrid 147
Rajan Singaravel, Hemachander Allamsetty, and Sankar Natarajan
5.1 Introduction 148
5.2 Microgrids 149
5.3 Drivers of Microgrid 150
5.3.1 Value drivers 150
5.3.2 Technology/policy drivers 151
5.4 Types of Microgrids 152
5.4.1 Classification of microgrids based on system topology 152
5.4.2 Classification of microgrids based on market segments 154
5.5 Components of Microgrid 155
5.5.1 Loads 155
5.5.2 Distributed energy resources (DERs) 156
5.5.3 Distributed energy storage (DES) 156
5.5.4 Microgrid control system 157
5.6 Hierarchical Control of MG 157
5.6.1 Conventional control hierarchy 157
5.6.1.1 Primary control 158
5.6.1.2 Secondary control 159
5.6.1.3 Tertiary control 159
5.6.2 Specialized hierarchical control architecture 159
5.6.2.1 Converter control level 160
5.6.2.2 Voltage coordination level 160
5.6.2.3 Energy management level 161
5.7 Methods of Control 161
5.7.1 Control strategies for primary control 161
5.7.1.1 Fuzzy logic control 161
5.7.1.2 DC bus signaling 162
5.7.1.3 Droop control 162
5.7.2 Secondary control scheme 162
5.7.3 Tertiary control 163
5.8 Conclusion 164

6 Multiport DC–DC Converter Fed Multilevel Inverter for Standalone Photovoltaic Systems 169
Hemachander Allamsetty, Lalit Kumar Sahu, and Praneeth Kumar Pedapati
6.1 Introduction 170

6.2 Operation of the Proposed Configuration 172
6.3 Simulation Results 174
6.4 Control Strategy 177
6.4.1 MPPT control 177
6.4.2 Inverter control 177
6.4.2.1 Generation of orthogonal phase 178
6.4.2.2 Control strategy 178
6.4.2.3 Controller design 178
6.5 Simulation Results 181
6.5.1 Case 1: Partial shading condition on one of the modules 181
6.5.2 Case 2: Dynamic loading scenarios 182
6.6 MPPT Efficiency Analysis 183
6.7 Conclusion 184

7 Smart Grid Communication and Protocols and Smart Interfacing of Electric Vehicles with the Grid 189
Kushan Tharuka, Ruvini De Seram, Anindita Golder, and Sheldon Williamson
7.1 Introduction 190
7.2 Advanced Metering Infrastructure (AMI) 191
7.2.1 Key components of AMI infrastructure 192
7.2.1.1 Smart metering 193
7.2.2 Implementation of AMI 195
7.2.2.1 Communication protocols 196
7.2.2.2 Radio frequency (RF) mesh 199
7.2.2.3 Power line communication (PLC) 200
7.2.2.4 Wi-Fi 201
7.2.2.5 LoRa and Low-Power Wide-Area (LPWA) 202
7.3 Smart Energy Management Systems 203
7.4 Power Electronic Converters for Grid Integration 207
7.4.1 EV-grid integration for V2G application 212
7.5 Conclusion 215

8 Implementation of V2G and G2V Methods with Distributed Generation for E-Mobility 219
Kodeeswaran S, Kannabhiran A, and Nandhini Gayathri M
8.1 Introduction 220

8.2 State of Electric Vehicle Market 221
8.2.1 EV market evolution 221
8.2.2 Support for charging infrastructure and policy incentives for electric vehicles 222
8.2.3 Renewable energy share in e-mobility markets . . . 223
8.2.4 EV flexibility potential 226
8.3 Smart Charging Outlook 226
8.3.1 Impact of charging EVs 226
8.3.2 Smart charging 227
8.3.3 Charging infrastructure 228
8.3.4 Smart charging enablers 229
8.4 Business Models and Regulatory Outlook 230
8.4.1 E-mobility performers 230
8.4.2 Business models and V2G connections 231
8.4.2.1 Smart energy services supplier and collector 231
8.4.2.2 Second-life storage applications 231
8.4.3 Guideline for V2G integration: Electrical energy markets 232
8.5 E-Mobility Outlook 232
8.5.1 Cost and competitiveness of EVs 232
8.5.2 Outlook for batteries 233
8.5.3 Shared e-mobility: Mobility as a service (MaaS) . . 234
8.5.4 Autonomous EVs 235
8.6 Smart Charging Influence on the Universal Energy System . 235
8.6.1 System-wide impact 235
8.6.2 Local distribution grid impact 236
8.7 Conclusion 236

Index **239**

About the Editors **241**

Preface

Electric Vehicles and Distributed Generation – Micro Grid

Although the cost of electricity has significantly decreased since the 1930s, it is still not readily available to the disadvantaged in countries and poverty regions. One of the single most ways to reduce poverty is to find ways to reduce the cost of energy. Renewable resources can play a significant role to reduce these costs. Electricity across our world is primarily produced from fossil-fuelled resources. There is substantial debate about the energy sources on today's fossil-fuelled power plants and the effects they may be having on the climate. A primary product of combustion of fossil generation is carbon dioxide, CO_2. Many of today's computer simulations suggest that as the atmospheric concentration of CO_2 increase, the earth's average temperature will continue to increase. The world is made up of countless ecosystems. Experimental investigations and field studies show that most ecosystems will collapse when subjected to fast temperature changes. To reduce the production of CO_2 globally, we must focus on getting the amount of CO_2 produced per unit of energy as close to zero as possible. As researchers, scientists, and engineers, we must focus on using our energy more efficiently and finding alternative energy sources that are competitive with fossil fuel energy sources. We must find ways to mitigate CO_2 production from our existing fossil fuel plants using technologies such as carbon capture and carbon storage in the near term. Wind, Solar photovoltaic (PV), and solar thermal are becoming cost-competitive but can be intermittent in nature and must be coupled through power electronics to existing energy systems. As the penetration of renewable resources increases, we must effectively use renewable energy produced through advanced load control strategies. Low cost, energy storage technologies will continue to develop and can abate renewable energy variability. Power electronics and their respective control systems are the enabling technology that will define our future energy.

The major advantage of the reduction in fossil fuel consumption, and greenhouse gas emission making hybrid vehicles, and electric vehicles more critical and famous in the automotive industry and as well as to the end-user.

In the past two decades, the advent of modern electric drives, battery technology, and intelligent charging methodology improved the overall performance of electric vehicles. Many researchers have been conducted to improve efficiency and reliability. It increases the applications of electric vehicles. Apart from grid power, the use of renewable sources makes zero-emission rate and zero fuel cost. The use of artificial intelligence introduces a significant improvement in electric vehicles' efficiency, reliability, and safety. A right technology means not only in the technical aspect, but also, it's economical. So it will reach an ordinary consumer also at an affordable price. It can be achieved by mean of using mass production by competitive manufacturers. This book aims to bring the fundamental concepts and a good understanding of vehicle electrification from the smart grid. This book will focus on Modern Wireless Electric Vehicle Drive Train Topologies, Energy Storage Systems, Wireless Charging Schemes, Power Converters and Power Quality Assessments in EV clusters. This book will enable researchers to focus their ideas to develop Real-time Hardware in Loop System for Wireless Charging of EV.

Chapter 1

As the world moves away from fossil fuels, the need for efficient, reliable, and innovative energy storage solutions has become more demanding than ever. This chapter highlights the various energy storage systems (ESS) technologies that are critical in the transition to electric vehicles (EVs) and hybrid electric vehicles (HEVs). It focuses on the advancements and applications of emerging technologies like lithium-ion batteries, solid-state batteries, fuel cells, and supercapacitors. It also emphasizes the importance of choosing the right energy storage solution based on performance needs, cost, safety, and technological progress, aiming to guide researchers, manufacturers, and EV developers in making informed decisions for sustainable transportation.

Chapter 3

The chapter summarizes the key findings and insights, as well as identifying the future directions and research opportunities for wireless charging technology for EVs. Also, this chapter aims to provide an in-depth understanding of the latest wireless charging schemes for EVs, and to highlight the potential of this technology in revolutionizing the EV industry.

Chapter 4

The instantaneous power theory of the multiphase power supply system was improved by substantiating new formulas for instantaneous active current and instantaneous apparent power which differ from the known ones by considering the different resistances of the transmission line. The strategy of shunt active filtering that forms this instantaneous active current provides the minimum of instantaneous power losses in the transmission line and surpasses filter strategies using pq-theory of instantaneous power and its modification, cross-vector theory, pqr-theory of instantaneous power, proportional-vector theory with deprivation of the zero-sequence voltage component.

A new formula for the active current vector of the multiphase power system based on modified integral apparent power resolution was proposed. Its realization by shunt active filter provides a minimum integral power loss in the transmission line and a unit value of the power factor. This filtering strategy outperforms other known filtering strategies based on active current formulas S. Fryze and P. Salmeron / J. Montano.

It has been found that the sinusoidal process in a three-phase four-wire power system with a symmetrical voltage source and asymmetric load is fully identified by six scalar values: active and reactive powers, two orthogonal components of the unbalance power and two orthogonal components of the power, associated with neutral wire current. These inactive powers can be used to synthesize control signals for active filtration in grid-tied distributed generation systems. Integral formulas were obtained to determine the mentioned inactive powers, making it possible to record their contribution to the deterioration of electrical energy quality.

The relationship between unbalanced power components and the linear unbalanced load parameters was established, which allowed us to verify these integral formulas by computer experiment. As has been shown these formulas preserve the output values of inactive powers in the presence of current higher harmonics.

Chapter 6

The integration of renewable energy sources, particularly photovoltaic (PV) systems, into power generation is a growing trend aimed at meeting the global demand for clean energy. However, the reliability and efficiency of these systems are challenged by factors such as partial shading, module mismatch, and the need for efficient power conversion systems. The work presented in this material addresses these critical issues by exploring advanced power conversion topologies designed for standalone PV applications.

Focusing on a multiport DC-DC converter-fed multilevel inverter configuration, this research highlights innovative solutions to mitigate power losses under varying operational conditions, including partial shading. The configuration, coupled with an efficient controller, ensures optimal performance by dynamically adapting to different PV panel inputs and regulating grid voltage.

This material offers valuable insights into the design, operation, and control strategies of next-generation transformerless PV systems, which are more reliable, cost-effective, and capable of handling dynamic environmental conditions. Through comprehensive simulation and analysis, this study advances the development of efficient power electronic systems for renewable energy applications, particularly in residential settings.

We hope this work will serve as a useful resource for researchers, engineers, and practitioners in the field of power electronics and renewable energy systems, further advancing the adoption and efficiency of photovoltaic technologies.

Chapter 7

The rapid integration of renewable energy systems (RES) and electric vehicles (EVs) has necessitated a transformation in modern power grids. The transition from traditional unidirectional power systems to advanced smart grids with bidirectional power flows has been discussed in this chapter. These smart grids offer a more efficient, reliable, and resilient electricity network by leveraging emerging technologies such as the Internet of Things (IoT), communication networks, cloud computing, machine learning, and advanced power electronics. Special emphasis is placed on integrating electric vehicles, which can consume and return energy to the grid through Vehicle-to-Grid (V2G) technology. This chapter also examines the challenges that come with this shift, including data privacy concerns, cybersecurity threats, and the complexities of managing advanced metering infrastructure.

Chapter 8

Implementation of V2G and G2V Methods with Distributed Generation for E-Mobility

This chapter addresses the need for smart charging infrastructure in electric vehicle (EV) systems, focusing on alternatives to traditional plug-in methods, which face issues related to bulky cables, safety, and continuous power supply. Wireless charging is presented as a solution to enhance user-friendliness and efficiency. However, with the increasing adoption of EVs, challenges arise for the power generation and distribution systems.

To tackle these issues, this chapter explores smart charging methods like Vehicle-to-Grid (V2G) and Grid-to-Vehicle (G2V) technologies, integrated with renewable energy sources (RES). These methods enable EVs to act as both energy consumers and providers, helping to balance grid demand and improve energy efficiency. The chapter also highlights the role of supporting software in calculating energy consumption and optimizing the bidirectional power flow between EVs and the grid.

In addition, this chapter reviews the infrastructure developments and policies adopted by various countries to support the growth of smart charging technologies. By examining these advancements, the chapter aims to provide a clear understanding of V2G and G2V integration, promoting future progress in the field of e-Mobility.

List of Figures

Figure 1.1	Basic structure of a fuel cell energy storage system.	8
Figure 1.2	Basic passive configuration.	25
Figure 1.3	Parallel SA-HESS.	27
Figure 1.4	Battery SA-HESS.	28
Figure 1.5	Capacitor SA-HESS.	29
Figure 1.6	Full-active HESS.	29
Figure 2.1	Classification of electric vehicle charging methods.	43
Figure 2.2	Three-phase on-board charger topology based on active rectifier and single DAB.	47
Figure 2.3	SRC DC–DC converter topology.	48
Figure 2.4	DAHB convertor topology.	49
Figure 2.5	Three-level NPC boost AC–DC converter topology.	50
Figure 2.6	Matrix converter topology with resonant tank and HF transformer for EV charger.	50
Figure 2.7	General block diagram of a single-phase on-board battery charger.	51
Figure 2.8	Unidirectional boost-based AC–DC converters.	52
Figure 2.9	Unidirectional interleaved boost-based AC–DC converters.	52
Figure 2.10	Bidirectional boost-based converters with synchronous rectifier.	53
Figure 2.11	Unidirectional nonisolated interleaved buck–boost DC–DC converters.	53
Figure 2.12	**(a)** Unidirectional isolated full-bridge series interleaved two quadrant buck DC–DC converters. **(b)** Bidirectional nonisolated resonant DC–DC converters.	53
Figure 2.13	Block diagram of a battery charger controller.	54
Figure 2.14	Frame work of EV with aggregators.	60
Figure 2.15	Vehicle-to-grid framework.	61
Figure 3.1	Schematic diagram of a typical MRWC system [7].	72

Figure 3.2 Schematic diagram of a typical MRWC system [7]. 73
Figure 3.3 General structure of an inductive wireless charging system for EVs [14]. 75
Figure 3.4 Schematic diagram of a typical CWC system. . . . 77
Figure 3.5 Efficiency of the system as a function of the distance between the charging pad and the receiver plate [16]. 78
Figure 3.6 Schematic diagram of a dynamic wireless charging system [19]. 80
Figure 3.7 Schematic diagram of an inductive power transfer-based wireless charging roadway system [21]. . . 82
Figure 4.1 Distributed generation system and consumption with centralized or distributed control. 93
Figure 4.2 Multiphase power supply system with resistive model of transmission line. 94
Figure 4.3 Geometric content of the solution of a partial optimization problem. 96
Figure 4.4 Three-phase four-wire power system using transmission line resistive model. 105
Figure 4.5 Dependence of the power factor on the integral voltage zero-sequence factor when applying the S. Fryze active filtering strategy. 113
Figure 4.6 Dependence of the power factor on the integral voltage zero-sequence factor when applying the P. Salmeron and J. Montano active filtering strategy. 114
Figure 4.7 Standardized model of three-phase four-wire power supply system. 115
Figure 4.8 Resistive model of three-phase four-wire power supply system in sinusoidal mode. 117
Figure 4.9 (a) Scheme for measuring power D_R.(b) Scheme for measuring power D_I. 126
Figure 4.10 Dependencies of powers from R_{AB} when $X_{AB}= 0$, and from X_{AB} when $R_{AB}= 5$. 127
Figure 4.11 The waveforms of instantaneous powers and their average values. 128
Figure 4.12 The load current waveform based on the calculated power component quantities. 129
Figure 4.13 Instantaneous load currents and power contributions to their rms values under linear load. 130

Figure 4.14 Instantaneous power components $p(t)$,$q(t)$,$d_R(t)$,$d_I(t)$ and their average values P, Q, D_R, D_I for linear and nonlinear load. 132
Figure 4.15 Instantaneous load currents and power contributions to their rms values under nonlinear load. . . 133
Figure 4.16 Instantaneous load currents and power contributions to their rms values in three-phase four-wire power system. 138
Figure 5.1 Simple diagram of a typical microgrid. 150
Figure 5.2 DC microgrid. 153
Figure 5.3 AC microgrid. 153
Figure 5.4 Hybrid AC–DC microgrid. 154
Figure 5.5 Conventional control architecture of microgrid. . . 158
Figure 5.6 Conventional control hierarchy with their major functions. 160
Figure 5.7 Classification of control algorithm. 164
Figure 6.1 Standard 1-phase transformerless converters: (a) one stage, (b) dual stage, (c) dual stage series configuration, (d) one stage series configuration, (e) dual stage parallel configuration, and (f) one stage parallel configuration. 171
Figure 6.2 Structure of the proposed dual input dual stage transformerless multilevel PV inverter. 172
Figure 6.3 Switching scheme for dual input structure. 173
Figure 6.4 Generation of switching signals using MC-SPWM. 174
Figure 6.5 Parameters of (a) panel 1, (b) panel 2, and (c) DC link and load. 175
Figure 6.6 Parameters of (a) panel 1, (b) panel 2, and (c) DC link and load. 176
Figure 6.7 Waveform of (a) leakage current and (b) common mode voltage. 177
Figure 6.8 Proposed structure with control strategy. 179
Figure 6.9 Parameters of (a) module 1 and (b) module 2. . . . 181
Figure 6.10 (a) Output voltage of MLI. (b) Grid voltage. (c) Current to the grid. 182
Figure 6.11 (a) Output voltage of MLI. (b) Grid voltage. (c) Current to the grid. 183
Figure 6.12 MPPT efficiency graph. 184

Figure 7.1 Smart grid architecture. 190
Figure 7.2 Closed-loop smart grid network architecture. . . . 192
Figure 7.3 Basic AMI system structure showing smart metering process. 193
Figure 7.4 Network architecture for smart meter. 195
Figure 7.5 Architecture diagram of the AMI system. 197
Figure 7.6 Wi-Fi network architecture 201
Figure 7.7 Energy management system framework. 204
Figure 7.8 Different devices to control the system voltage. . . 205
Figure 7.9 System frequency change and response to control action. 206
Figure 7.10 Types of power electronic converters. 207
Figure 7.11 DC–AC converter diagrams 208
Figure 7.12 Main types of DC–DC converters. 209
Figure 7.13 Three-phase AC–AC converter. 209
Figure 7.14 Sample architecture of a grid-integrated renewable system. 211
Figure 7.15 Controlling of a grid-tied inverter for PV integration. 212
Figure 7.16 V2G application. 212
Figure 7.17 An overview of energy management of V2G. . . . 214
Figure 8.1 Country-/region-wise total electric vehicle sales up to 2019. 221
Figure 8.2 Block diagram of wind and solar interface. 225
Figure 8.3 Renewable energy generation by countries. 225
Figure 8.4 Smart charging formation. 228
Figure 8.5 EV wireless charging infrastructure. 229
Figure 8.6 Real average data for Li-ion batteries. 233
Figure 8.7 Chain diagram for MaaS. 234

List of Tables

Table 1.1 Electrochemical reactions of lithium nickel manganese cobalt oxide (NMC). battery. 4
Table 1.2 Electrochemical reactions of lithium nickel cobalt aluminum (NCA) oxide battery. 4
Table 1.3 Electrochemical reactions of lithium iron phosphate (LFP) battery. 5
Table 1.4 Electrochemical reactions of lithium manganese oxide (LMO) battery. 6
Table 1.5 Electrochemical reactions of lithium titanium oxide (LTO) battery. 7
Table 1.6 Electrochemical reactions of fuel cell. 7
Table 1.7 Types and Characteristics of fuel cells [1]. 9
Table 1.8 Electrochemical reactions of lithium-sulfur battery. . 11
Table 1.9 Electrochemical reactions of nickel-metal hydride (NiMH) battery. 11
Table 1.10 Performance comparison of supercapacitor and lithium-ion battery. 12
Table 1.11 Advantages and disadvantages of automotive grade supercapacitors. 14
Table 1.12 Electrochemical reactions of flow battery. 18
Table 3.1 Different wireless charging schemes for electric vehicles (EVs). 84
Table 6.1 Switching states of T-type MLI. 173
Table 6.2 Switching paths of T-type MLI. 174
Table 6.3 Parameters used. 175
Table 6.4 Parameters of the module. 181
Table 7.1 ISM band frequency. 200
Table 8.1 E-mobility targets and policy for different countries. 223
Table 8.2 EVs charging infrastructure implementation. 224
Table 8.3 Mobility actors positioning. 231

List of Contributors

A., Kannabhiran, *School of Electrical and Electronics Engineering, SASTRA Deemed to be University, India; E-mail: mail2akannan@gmail.com*

Allamsetty, Hemachander, *Department of Electrical and Electronics Engineering, National Institute of Technology Puducherry, India; E-mail: hemachander.a@nitpy.ac.in*

Anekal, Latha, *Smart Transportation Electrification and Energy Research (STEER) Group Department of Electrical, Computer, and Software Engineering Ontario Tech University Oshawa, Canada; E-mail: latha.anekal@ontariotechu.net*

Artemenko, M. Yu., *Department of Acoustic and Multimedia Electronic Systems of the National Technical University of Ukraine, Igor Sikorsky Kyiv Polytechnic Institute, Ukraine; E-mail: artemenko_m_ju@ukr.net*

Arul, Asis Allwyn Clarence, *Department of Electrical and Electronics Engineering, Mepco Schlenk Engineering College, India; E-mail: iamallwyn@mepcoeng.ac.in*

Chetri, Chandan, *Smart Transportation Electrification and Energy Research (STEER) Group Department of Electrical, Computer, and Software Engineering Ontario Tech University Oshawa, Canada; E-mail: chandan.chetri@ontariotechu.net*

Chopyk, V. V., *Institute of Electrodynamics of the National Academy of Sciences of Ukraine, Ukraine; E-mail: chopykied@gmail.com*

De Seram, Ruvini, *Smart Transportation Electrification and Energy Research(STEER) Group Department of Electrical, Computer, and Software Engineering Ontario Tech University Oshawa, Canada; E-mail: edirisooriyamohottige.deseram@ontariotechu.net*

Golder, Anindita, *Smart Transportation Electrification and Energy Research(STEER) Group Department of Electrical, Computer, and Software Engineering Ontario Tech University Oshawa, Canada; E-mail: anindita.golder@ontariotechu.net*

Iqteit, Nassim A., *Department of Electrical Engineering, Palestine Polytechnic University, Palestine; E-mail: nassim_eng83@ppu.edu*

Jayakumar, Roscia Jeya Shiney, *Department of Electronics and Communication Engineering, P.S.R Engineering College, India; E-mail: shineyroscia@gmail.com*

M., Nandhini Gayathri, *School of Electrical and Electronics Engineering, SASTRA Deemed to be University, India; E-mail: nandhini.gayathri@gmail.com*

Mykhalskyi, V. M., *Institute of Electrodynamics of the National Academy of Sciences of Ukraine, Ukraine; E-mail: mikhalsky@ied.org.ua*

Natarajan, Sankar, *Department of Electrical and Electronics Engineering, National Institute of Technology Puducherry, India; E-mail: naseee2591@gmail.com*

Pedapati, Praneeth Kumar, *Department of Electrical & Electronics Engineering, NIT Puducherry, India; E-mail: ee21d1008@nitpy.ac.in*

Prabaharan, N., *Department of Electrical & Electronics Engineering, SASTRA Deemed University, India; E-mail: prabaharan.nataraj@gmail.com*

S., Kodeeswaran, *Department of Electrical Engineering, National Institute of Technology, Andhra Pradesh, India; E-mail: kodikrct@gmail.com*

Sahu, Lalit Kumar, *Department of Electrical Engineering, NIT Raipur, India; E-mail: lkumar.ele@nitrr.ac.in*

Salem, Mohamed, *School of Electrical and Electronic Engineering, Universiti Sains Malaysia (USM), Malaysia; E-mail: salemm@usm.my*

Samanta, Akash, *Smart Transportation Electrification and Energy Research (STEER) Group Department of Electrical, Computer, and Software Engineering Ontario Tech University Oshawa, Canada E-mail: akash.samanta@ontariotechu.net*

Shapoval, I. A., *Institute of Electrodynamics of the National Academy of Sciences of Ukraine, Ukraine; E-mail: shapoval@ied.org.ua*

Singaravel, Rajan, *Department of Electrical and Electronics Engineering, National Institute of Technology Puducherry, India; E-mail: rajan.singaravel@nitpy.ac.in*

Tharuka, Kushan, *Smart Transportation Electrification and Energy Research(STEER) Group Department of Electrical, Computer, and Software Engineering Ontario Tech University Oshawa, Canada; E-mail: kushan.lulbaddawaduge@ontariotechu.net*

Williamson, Sheldon, *Smart Transportation Electrification and Energy Research (STEER) Group Department of Electrical, Computer, and Software Engineering Ontario Tech University Oshawa, Canada; E-mail: sheldon.williamson@ontariotechu.ca*

Yahya, Adel E. M., *Department of Electrical Engineering, Regdalen, Sabratah University, Libya; E-mail: ecc.yahya@gmail.com*

Yahya, Khalid, *Department of Electrical and Electronics Engineering, Faculty of Engineering and Architecture, Istanbul Gelisim University, Istanbul 34310, Turkey; E-mail: hayahya@gelisim.edu.tr*

List of Abbreviations

ACE	Area control error
AHES	Active hybrid energy storage
AIB	Aluminum-ion battery
AMI	Advanced metering infrastructure
AP	Application provider
APD	Active power dispatch
ARRA	American Recovery and Reinvestment Act
AT&C	Aggregate technical and commercial
BAU	Business as usual
BB	Broadband
BCR	Battery current regulation
BEV	Battery-operated vehicle
BMS	Battery management system
BS	Base station
BSA	Battery semi-active
BSS	Battery swap station
CAES	Compressed air energy storage
CBSS	Concealed based security scheme
CC	Constant current
CHP	Combined heat and power
CISPR	International special committee on radio interference
CSA-HESS	Capacitor semi-active hybrid energy storage system
CSC	Current source converter
CV	Constant voltage
CWC	Capacitive wireless charging
DAB	Dual active bridge
DAHB	Dual active half bridge
DER	Distributed energy resource
DES	Distributed energy storage

DG	Distributed generation
DWC	Dynamic wireless charging
EDLC	Electrical double-layer capacitor
EMC	Electromagnetic compatibility
EMI	Electromagnetic inference
EMS	Energy management systems
ESS	Energy storage system
EV	Electric vehicle
EVSE	Electric vehicle supply equipment
FACTS	Flexible AC transmission systems
FA-HES	Full-active hybrid energy storage
FCEV	Fuel cell electric vehicle
FCHEV	Fuel cell hybrid electric vehicle
GA	Genetic algorithm
GPU	Graphics processing unit
HASS	Hybridized energy accumulation system
HESS	Hybrid energy storage system
HEV	Hybrid electric vehicle
HVAC	Heating, ventilation, and air conditioning
HVDC	High-voltage direct current
IBR	Inverter-based resource
ICE	Internal combustion engine
IEA	International Energy Agency
IEC	International Electrotechnical Commission
IoT	Internet of things
IP	Internet portal
IPT	Inductive power transfer
IRENA	International Renewable Energy Agency
ITU-R	International Telecommunication Union Radiocommunication
LDC	Linearly declining current
LFP	Lithium iron phosphate
LFR	Loss-free resistor
LIB	Lithium-ion battery
LMO	Lithium manganese oxide
LMP	Lithium metal polymer
LoS	Line of sight
LTO	Lithium titanium oxide
MaaS	Mobility as a service

MDAS	Meter data acquisition system
MDM	Meter data management
MDMS	Master data management system
MDP	Markov decision process
MG	Microgrid
MLI	Multilevel inverter
MMFA-HES	Modular multilevel full-active hybrid energy storage
MPPT	Maximum power point tracking
MRWC	Magnetic resonance-based wireless charging
NAN	Neighborhood area network
NB	Narrowband
NCA	Nickel cobalt aluminum
NiMH	Nickel-metal hydride
NMC	Nickel manganese cobalt
NPC	Neutral point clamped
OEM	Original equipment manufacturer
OFC	Optical fiber communication
OFDM	Orthogonal frequency division multiplexing
P&O	Perturb and observe
PCM	Phase change material
PEM	Proton exchange membrane
PEMFC	Proton-conducting (acidic) electrolytes fuel cell
PEV	Pure electric vehicle
PFC	Power factor correction
P-HASS	Passive hybridized energy storage system
PHEV	Plug-in hybrid electric vehicle
P-HS	Passive hybrid storage
PI	Proportional-integral
PLC	Power line communication
PLCC	Power-line carrier communication
POPI	power output = power input
PSA-HES	Semi-active and partially parallel hybrid energy storage
PWM	Pulse width modulation
REEV	Range extended electric vehicle
RES	Renewable energy source
RF	Radio frequency
RMS	Root mean square

RPD	Reactive power dispatch
SAE	Society of Automotive Engineers
SAF	Shunt active filter
SA-HES	Semi-active hybrid energy storage
SIB	Sodium-ion battery
SM	Smart meter
SMC	Sliding mode control
SOC	State of charge
SRC	Series-resonant converter
SRF	Synchronous reference frame
SSB	Solid-state battery
SSN	Silver Spring Network
T&D	Transmission and distribution
TCP	Transmission control protocol
TEG	Thermoelectric generator
TESS	Thermal energy storage system
THD	Total harmonic distortion
TMS	Thermal management strategy
ToD	Time of day
UART	Universal asynchronous receiver-transmitter
UP	Utility provider
UPS	Uninterruptible power supplies
V2B	Vehicle to building
V2G	Vehicle to grid
V2H	Vehicle to home
V2X	Vehicle to everything
VFD	Variable frequency drive
VRFB	Vanadium redox flow battery
VSC	Voltage source converter
WAN	Wide area network
WPT	Wireless power transfer
WS	Web server
ZEBR	A Zeolite Battery Research Africa
ZVS	Zero voltage switching

1

Energy Storage Systems for Transportation Electrification

Chandan Chetri, Akash Samanta, Latha Anekal, and Sheldon Williamson

Smart Transportation Electrification and Energy Research (STEER) Group
Department of Electrical, Computer, and Software Engineering Ontario Tech University Oshawa, Canada
E-mail: chandan.chetri@ontariotechu.net; akash.samanta@ontariotechu.net; latha.anekal@ontariotechu.net; sheldon.williamson@ontariotechu.ca

Abstract

This chapter focuses on energy storage systems (ESSs) and their growing significance in transforming the transportation sector through electrification. As the global shift toward sustainable and eco-friendly mobility intensifies, the demand for efficient, reliable, and innovative energy storage solutions has become essential. A wide range of ESSs are used in the transportation industry. Earlier mostly lead-acid batteries are used for auxiliary and emergency power supply in conventional gasoline cars. Recently, especially in the last two decades, lithium-ion batteries (LIBs) have been widely used in fully electric vehicles (EVs) and hybrid electric vehicles (HEVs), primarily due to their high energy density and relatively long cycle life. The chapter provides an extensive review of existing and emerging technologies, offering a comprehensive understanding of their functionalities, applications, and potential implications for the future of transportation. The choice of an optimum energy storage system for an EV depends on several factors and the requirements of customers. Therefore, a detailed discussion of each of the energy storage technologies is discussed in this chapter. The primary aim of this chapter is to provide a ready reference to the researchers, battery manufacturers, and the EV original equipment manufacturer to make a wise decision regarding the selection of a suitable energy storage system for a particular EV. It is noteworthy that the integration of these technologies in transportation is dependent upon various aspects, including but not limited to cost, energy density, safety, and ongoing technological advancements. The field is dynamic, with continuous research and development. The book chapter concludes by addressing the broader landscape and discussing the challenges, opportunities, and future directions in energy storage for transportation electrification. By examining the multifaceted nature of ESSs, this chapter aims to equip researchers, practitioners, and enthusiasts with a comprehensive understanding of the evolving field, fostering advancements that drive sustainable and efficient transportation solutions for the future.

Keywords: Energy Storage Systems (ESS), Lithium-ion Battery (LIB), Electric Vehicle (EV), Battery Management Systems (BMSs).

1.1 Introduction to Energy Storage System for EVs

Energy storage systems (ESSs) for electric vehicles (EVs) and hybrid electric vehicles (HEVs) have received major attention from the transportation community, especially in the last two decades. The proliferation of fully electric, hybrid, and plug-in EVs together with the novel concept of electrified highways are the primary reasons behind the increasing interest in ESSs [1]. Primarily lithium-ion batteries (LIBs) have been widely used in EVs recently [2]. There are several ESSs available that have significant potential to be used in transportation electrification which include: solid-state batteries [3], nickel-metal hydride (NiMH) batteries [4], lithium–sulfur batteries [5], hydrogen fuel cells [6], supercapacitors [7], flywheel energy storage [8], thermal energy storage [9], graphene-based batteries [10], flow batteries [11], aluminum-ion batteries [12], zinc-air batteries [13], sodium-ion batteries [14], molten salt batteries [15], compressed air energy storage (CAES) [16], thermal electric generators [17]. Beginning with thorough exploration of LIBs, the backbone of contemporary EVs, the chapter navigates through advancements in solid-state batteries, promising higher energy densities and enhanced safety profiles. Hydrogen fuel cells, heralding a new era of zero-emission transportation, are discussed in detail, shedding light on their applications in fuel cell electric vehicles (FCEVs) and their unique advantages. The chapter also discusses unconventional storage methods, including supercapacitors, flywheel energy storage, and thermal energy storage. These technologies, while niche, contribute significantly to specific aspects of transportation electrification, such as regenerative braking and thermal management. Furthermore, the potential of emerging technologies like graphene-based batteries, aluminum-ion batteries, and sodium-ion batteries is explored, providing insights into their unique characteristics and ongoing research efforts. The requirement of ESSs significantly varies from one transportation application to the other, making it difficult for the scientific community to choose among the available options. This chapter reviews the available storage technology, the key performance demands of the major transportation applications, current issues and challenges, and suggestions for future improvement and direction to choose an optimum energy storage system for different application needs.

1.2 Widely used Energy Storage Systems

1.2.1 Lithium-ion battery

Lithium-ion batteries are rechargeable energy storage devices widely used in modern technology for their high energy density and lightweight design. They operate by moving lithium ions between positive and negative electrodes during charge and discharge cycles. These batteries are commonly found in portable electronics, EVs, and grid ESSs.

1.2.1.1 Lithium nickel manganese cobalt oxide (NMC) battery

Lithium nickel manganese cobalt oxide (NMC) batteries have become a major advancement in energy storage technology, particularly for applications such as EVs. NMC batteries feature a cathode composed of a mixture of lithium, nickel, manganese, and cobalt oxides, offering a delicate balance between energy density, power output, and thermal stability. The cathode composition is typically expressed as $LiNi_xMn_yCo_zO_2$, where x, y, and z represent the respective molar proportions of nickel, manganese, and cobalt which add up to 1 [18]. The anode, usually made of graphite, serves as a host for lithium ions during charge and discharge cycles, while the electrolyte facilitates the movement of lithium ions between the cathode and anode. The working principle of NMC batteries involves the intercalation and deintercalation of lithium ions within both the cathode and anode throughout the charging and discharging cycles. The comprehensive electrochemical reaction within the NMC battery is depicted in Table 1.1, illustrating that the discharge process advances from left to right, while the charging reaction progresses from right to left [19].

Table 1.1 Electrochemical reactions of lithium nickel manganese cobalt oxide (NMC). battery.

Electrode	Electrochemical reactions
Anode	$Li_nC_6 \rightleftharpoons Li_0C_6 + nLi^+ + ne^-$
Cathode	$Li_{m-n}(Ni_xMn_yCo_z)O_2 + nLi^+ + ne^- \rightleftharpoons Li_m(Ni_xMn_yCo_z)O_2$
Overall reaction	$Li_nC_6 + Li_{m-n}(Ni_xMn_yCo_z)O_2 \rightleftharpoons Li_0C_6 + Li_m(Ni_xMn_yCo_z)O_2$

NMC batteries typically have a nominal voltage ranging from 3.6 to 3.7 volts per cell and an energy density ranging from 150 to 200 Wh/kg. These characteristics make NMC batteries well-suited for EV applications, where high energy density and power output are crucial for achieving long driving ranges and supporting fast-charging capabilities.

1.2.1.2 Lithium nickel cobalt aluminum (NCA) oxide

Lithium nickel cobalt aluminum (NCA) batteries are a prominent type of LIB utilized in various applications. NCA batteries substitute aluminum for manganese in NMC batteries, enhancing both specific energy and lifespan compared to the NMC variant [20]. The typical composition of an NCA cathode is expressed as $LiNi_xCo_yAl_zO_2$, where *x*, *y*, and *z* represent the respective molar proportions of nickel, cobalt, and aluminum and add up to 1 [21].

The working principle of NCA batteries involves the intercalation and deintercalation of lithium ions within both the cathode and anode throughout the charging and discharging cycles. The comprehensive electrochemical reaction within the NCA battery is depicted in Table 1.2, illustrating that the discharge process advances from left to right, while the charging reaction progresses from right to left [19].

NCA batteries are renowned for their high energy density, typically ranging from 250 to 300 Wh/kg, and a nominal voltage of 3.6 to 3.7 volts per cell. These characteristics make NCA batteries highly suitable for EV applications, where energy density and power output are crucial factors for achieving long driving ranges and supporting fast-charging capabilities. Additionally,

Table 1.2 Electrochemical reactions of lithium nickel cobalt aluminum (NCA) oxide battery.

Electrode	Electrochemical reactions
Anode	$Li_nC_6 \rightleftharpoons Li_0C_6 + nLi^+ + ne^-$
Cathode	$Li_{m-n}(Ni_xCo_yAl_z)O_2 + nLi^+ + ne^- \rightleftharpoons Li_m(Ni_xCo_yAl_z)O_2$
Overall reaction	$Li_nC_6 + Li_{m-n}(Ni_xCo_yAl_z)O_2 \rightleftharpoons Li_0C_6 + Li_m(Ni_xCo_yAl_z)O_2$

Table 1.3 Electrochemical reactions of lithium iron phosphate (LFP) battery.

Electrode	Electrochemical reactions
Anode	$Li_nC_6 \rightleftharpoons Li_0C_6 + nLi^+ + ne^-$
Cathode	$Li_{m-n}FePO_4 + nLi^+ + ne^- \rightleftharpoons Li_mFePO_4$
Overall reaction	$Li_nC_6 + Li_{m-n}FePO_4 \rightleftharpoons Li_0C_6 + Li_mFePO_4$

NCA batteries offer excellent thermal stability and cycle life, contributing to their reliability and longevity in demanding automotive environments.

1.2.1.3 Lithium iron phosphate (LFP) battery

Lithium iron phosphate (LFP) batteries represent a notable subset of LIBs renowned for their remarkable safety, long cycle life, and thermal stability, making them a preferred choice for various applications, particularly in EVs. These batteries feature a cathode composed of lithium iron phosphate, a compound known for its stable crystal structure and robust performance characteristics [22]. The anode, typically constructed from graphite or other carbon-based materials, serves as a host for lithium ions during charge and discharge cycles, while the electrolyte facilitates the movement of lithium ions between the cathode and anode. The typical composition of an LFP cathode is expressed as $LiFePO_2$ [23]. The operational principle of LFP batteries revolves around the intercalation and deintercalation of lithium ions within the cathode and anode during the charging and discharging cycles. The comprehensive electrochemical reaction within the LFP battery is depicted in Table 1.3, illustrating that the discharge process advances from left to right. In contrast, the charging reaction progresses from right to left [19].

$LiFePO_4$ batteries typically exhibit a nominal voltage of around 3.2 volts per cell and energy density ranging from 90 to 160 Wh/kg. Although $LiFePO_4$ batteries have lower energy density than other LIB chemistries, they excel in safety, thermal stability, and cycle life. LFP batteries are commonly used in electric buses and commercial vehicles due to their robustness, safety, and relatively lower cost compared to other lithium-ion chemistries.

1.2.1.4 Lithium manganese oxide (LMO) battery

Lithium manganese oxide (LMO) batteries utilize manganese oxide as the cathode material. LMO batteries are renowned for their safety, stability, and cost-effectiveness, making it a popular choice for various applications [24]. The cathode of an LMO battery typically consists of lithium manganese oxide ($LiMn_2O_4$), while the anode is commonly composed of graphite [25]. The

Table 1.4 Electrochemical reactions of lithium manganese oxide (LMO) battery.

Electrode	Electrochemical reactions
Anode	$Li_nC_6 \rightleftharpoons Li_0C_6 + nLi^+ + ne^-$
Cathode	$Li_{m-n}Mn_2O_4 + nLi^+ + ne^- \rightleftharpoons Li_mMn_2O_4$
Overall reaction	$Li_nC_6 + Li_{m-n}Mn_2O_4 \rightleftharpoons Li_0C_6 + Li_mMn_2O_4$

operational principle of LMO batteries comprises of the intercalation and deintercalation of lithium ions within the cathode and anode during charging and discharging cycles. The comprehensive electrochemical reaction within the LFP battery is depicted in Table 1.4, illustrating that the discharge process advances from left to right, while the charging reaction progresses from right to left [19].

The nominal voltage of LMO batteries typically ranges from 3.6 to 3.8 volts per cell and offers energy densities suitable for 100 to 150 Wh/kg. LMO batteries find particular applications in HEVs and plug-in hybrid electric vehicles (PHEVs), where reliable energy storage solutions are provided and low cost for manufacture, which are important considerations in automotive applications. Moreover, LMO batteries are often combined with other battery chemistries or used as part of battery packs for better performance and meet the specific needs of EV manufacturers.

1.2.1.5 Lithium titanium oxide (LTO) battery

Lithium titanium oxide (LTO) is a promising material used in LIBs renowned for its exceptional safety, longer cycle life, and fast-charging capabilities, making it an ideal choice for demanding applications such as EVs [26]. Instead of using traditional carbon-based anodes, LTO batteries employ lithium titanate nanocrystals, a spinel structure compound with the chemical formula $Li_4Ti_5O_{12}$, where lithium ions are incorporated within the lattice structure of titanium oxide [27]. This unique structure allows for the rapid insertion and extraction of lithium ions during charge and discharge cycles, facilitating fast charging and discharging without compromising battery performance or safety. The working principle of LTO batteries comprises the reversible intercalation and deintercalation of lithium ions within the crystal lattice of titanium oxide during charging and discharging cycles. The anode and cathode reactions are shown in Table 1.5 [28]. The specific reactions depend on the composition of the electrode materials and battery operating conditions.

LTO batteries typically exhibit a nominal voltage of around 2.4 volts per cell and a lower energy density compared to other LIB chemistries, typically

Table 1.5 Electrochemical reactions of lithium titanium oxide (LTO) battery.

Electrode	**Electrochemical reactions**
Anode	$Li_7Ti_5O_{12} \rightleftharpoons Li_4Ti_5O_{12} + 3Li^+ + 3e^-$
Cathode	$Li_{m-n}Mn_2O_4 + nLi^+ + ne^- \rightleftharpoons Li_mMn_2O_4$
Overall reaction	$Li_7Ti_5O_{12} + Li_{1-x}C_6 \rightleftharpoons Li_4Ti_5O_{12} + LiC_6$

ranging from 60 to 110 Wh/kg. Despite the lower energy density, LTO batteries undergo minimal chemical changes during charging and discharging cycles, providing several advantages, including exceptional safety, long cycle life exceeding 6000 cycles, and the ability to operate over a wide temperature range. LTO batteries are well-suited for applications requiring fast-charging capabilities, such as public transportation and heavy-duty vehicles.

1.3 New and Potential Future Use Energy Storage Systems

1.3.1 Fuel cell

Fuel cells are electrochemical devices that generate electrical energy by converting the stored energy into fuels, such as hydrogen. They work by breaking down the fuel into its constituent parts (hydrogen and oxygen) and then recombining them in a controlled manner to produce electrical energy, water, and heat. When the supplied hydrogen gas encounters the anode, then hydrogen molecule gets split into hydrogen ion and electrons as shown in the equation given below in Table 1.6. These hydrogen ions travel through the electrolyte and reach the cathode, where it gets combined with oxygen molecules to form water and release heat, as shown in the equation given below in Table 1.6. Figure 1.1 shows the principle of operation of fuel cells. In a fuel cell, gaseous fuel such as hydrogen is fed continuously to the negative electrode (anode) and oxygenated air is fed continuously to the positive electrode (cathode). The product of the fuel cell are electricity, water, and heat, obtained from the electrochemical reactions. The comprehensive electrochemical reaction within a fuel cell is depicted in Table 1.6, illustrating that the discharge process advances from left to right, while the fuel cell acts as an electrolyzer from right to left [29].

Table 1.6 Electrochemical reactions of fuel cell.

Electrode	**Electrochemical reactions**
Anode	$H_2 \rightleftharpoons 2H^+ + 2e^-$
Cathode	$\frac{1}{2}O_2 + 2H^+ + 2e^- \rightleftharpoons H_2O$
Overall reaction	$H_2 + \frac{1}{2}O_2 \rightleftharpoons H_2O$

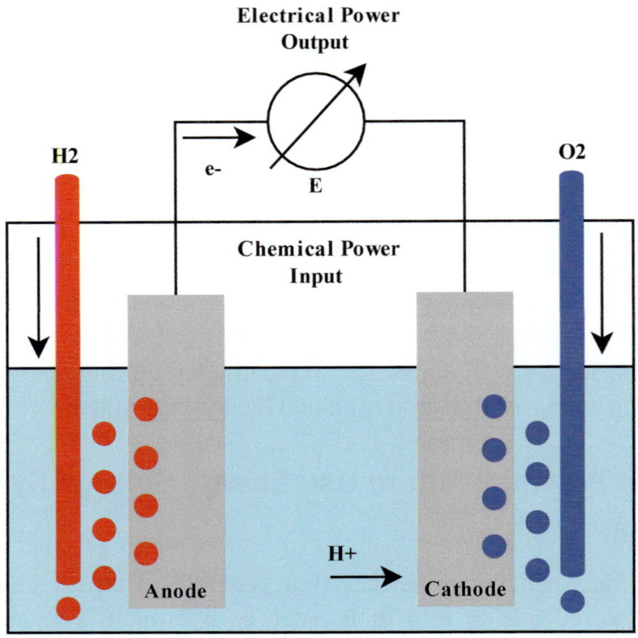

Figure 1.1 Basic structure of a fuel cell energy storage system.

Fuel cells are continuously provided with reactants from an external source. It is an energy conversion device that functions continuously without the chance of getting exhausted if fuel and oxidant are supplied to its electrodes. The provided table, labeled as Table 1.7, offers a comprehensive overview of the different types of fuel cells commonly used today [1]. Fuel cells are categorized based on the type of ion conductor or electrolyte they utilize. One important category consists of fuel cells that employ proton-conducting (acidic) electrolytes fuel cells (PEMFCs). These electrolytes can be in the form of a solid membrane, or a liquid acid absorbed in a polymer matrix such as phosphoric acid fuel cells. Another category of fuel cells operating at very high temperatures are either oxidic solids that conduct O^{2-} ions or molten carbonates CO_3^{2-} that conduct ions as their electrolytes. The electrolyte and temperature determine the choice of fuel and catalysts. While high-temperature fuel cells may run on a larger variety of fuels, low-temperature fuel cells typically employ hydrogen or liquid organic molecules. Another category comprises fuel cells that use hydroxide-conducting electrolytes which are alkaline. They can be in the form of a solid membrane (alkaline membrane fuel cells) or a liquid electrolyte (alkaline

Table 1.7 Types and Characteristics of fuel cells [1].

Fuel cell type	Electrolyte	Charge career	Operating temperature	Fuel	Efficiency
Proton exchange	Solid polymer	H^+	50 °C–100 °C	Pure H_2	35%–45%
Phosphoric acid	Phosphoric acid	H^+	~220 °C	Pure H_2	40%
Alkaline	KOH	OH^-	60 °C–120 °C	Pure H_2	35%–55%
Alkaline membrane	Solid polymer	OH^-	50 °C–100 °C	Pure H_2	<35%
Solid oxide	Solid oxide electrolyte	O^{2-}	~1000 °C	H_2, COCO, CH_4, and other hydrocarbons	>50%
Molten carbonate	Solid oxide electrolyte	CO_3^{2-}	~650 °C	H_2, COCO, CH_4, and other hydrocarbons	>50%

fuel cells). Acidic electrolytes generally require more expensive noble metal electrocatalysts, while alkaline fuel cells can tolerate less expensive catalysts. High-temperature fuel cells often incorporate oxidic electrocatalytic materials.

One of the biggest advantages of fuel cells is their high efficiency due to their ability to produce electricity directly from the supplied fuel. Unlike traditional combustion engines, fuel cells do not produce any emissions, making them a cleaner and more sustainable source of energy. Fuel cells are also relatively quiet and have a long lifespan, making them ideal for stationary and mobile applications. As a single fuel cell produces very low voltage, so these can be stacked up in series to increase the power level from few watts to several thousand kilowatts. Fuel cell ESSs come in different types, including alkaline, proton exchange membrane (PEM), solid oxide, and molten carbonate fuel cells. Each type of fuel cell possesses distinct advantages and disadvantages, and its selection is dependent upon particular application and specifications. PEM fuel cells are most commonly used in ESSs. They are lightweight, compact, and have a fast start-up time, making them ideal for mobile applications such as EVs. On the other hand, solid oxide fuel cells find optimal use in stationary applications such as backup power for homes and businesses. They are also capable of running on a variety of fuels, including natural gas, biogas, and hydrogen. One of the major drawbacks of fuel cell ESSs is its cost. The cost of the fuel cell system depends on the type of fuel cell used, the size of the system, and the application. However, it is anticipated that the cost of fuel cell technology will lessen with ongoing research and development efforts, and as the technology becomes more widespread. In conclusion, fuel cell ESSs are a promising technology that offers many advantages over traditional ESSs. They are clean, efficient, and can be used for a variety of applications. However, their high cost remains a major barrier to widespread adoption, and more research and development is needed to improve their cost-effectiveness.

1.3.2 Solid-state battery

Solid-state batteries (SSBs) represent the next advancement in battery technology, aiming to replace the traditional liquid electrolyte with a solid-state electrolyte. This system eliminates the risks associated with flammable electrolytes, increases stability, increases energy density, and improves longevity [30]. SSBs typically use lithium metal as the anode, a solid electrolyte such as lithium garnet, polyvinylidene fluoride, and a cathode material like LFP or NMC. During the charging and discharging cycles, lithium ions migrate between the anode and cathode through the solid electrolyte, facilitating the flow of electric current. SSBs hold immense potential for EVs due to their higher energy density, improved safety profile, and the potential for fast-charging applications [31]. By replacing the combustible liquid electrolyte with a solid-state equivalent, these batteries mitigate the risk of thermal runaway and improve the overall safety of EVs. Additionally, their higher energy density can extend the driving range of EVs, while faster charging times can reduce downtime for recharging.

1.3.3 Lithium-sulfur battery

Lithium-sulfur (Li-S) batteries have emerged as promising candidates for next-generation ESSs due to their high theoretical energy density and low cost [32]. Li-S batteries operate based on the reversible electrochemical reactions between lithium ions and sulfur. During discharge, lithium ions migrate from the lithium metal anode to the sulfur cathode, where they react with sulfur to form lithium sulfide (Li_2S). This reaction produces high energy density ranging from 350 to 609 Wh/kg [33]. During the process of charging, the lithium ions move back to the anode. The comprehensive electrochemical reaction within the Li-S battery is depicted in Table 1.8, illustrating that the discharge process advances from left to right, while the charging reaction progresses from right to left [34].

Li-S batteries provide a means of enabling long-range EVs with reduced weight and cost compared to conventional LIBs. Their high energy density and potential for lower manufacturing costs make them advantageous for

Table 1.8 Electrochemical reactions of lithium-sulfur battery.

Electrode	Electrochemical reactions
Anode	$16Li \rightleftharpoons 16Li^{+} + 16e^{-}$
Cathode	$S_8 + 16Li^{+} + 16e^{-} \rightleftharpoons 8\text{Li}_2S$
Overall reaction	$16Li + S_8 \rightleftharpoons 8Li_2S$

EV manufacturers seeking to extend their driving range. However, further research and development are necessary to overcome technical difficulties and commercialize Li-S batteries for widespread adoption in the automotive industry.

1.3.4 Nickel-metal hydride (NiMH) battery

Nickel-metal hydride (NiMH) batteries represent a prominent class of rechargeable energy storage devices that have garnered considerable attention due to their versatility and efficiency [35]. NiMH batteries are composed of a nickel oxide hydroxide (NiOOH) cathode and a hydrogen-absorbing alloy anode, NiMH batteries operate on the principle of reversible electrochemical reactions. While charging, a voltage is applied across the battery terminals, inducing oxidation of the nickel hydroxide within the positive electrode. This process releases oxygen ions and electrons. Concurrently, the hydrogen-absorbing alloy in the negative electrode undergoes oxidation, forming metal hydrides by absorbing hydrogen ions and electrons. This process enables the storage of energy within the metal hydride lattice through chemical bonds. Upon discharging, the nickel hydroxide is reduced back to its original state, releasing stored electrons, while the metal hydride oxidizes, releasing stored hydrogen ions and electrons. The anode and cathode reactions are shown in Table 1.9, illustrating that the charge process advances from left to right, while the discharge reaction progresses from right to left [36].

The efficient and reversible nature of these electrochemical reactions enables NiMH batteries to exhibit higher energy densities, fast-charging capabilities, and extended cycle life compared to other rechargeable battery technologies. These characteristics make NiMH batteries highly suitable for

Table 1.9 Electrochemical reactions of nickel-metal hydride (NiMH) battery.

Electrode	Electrochemical reactions
Anode	$M + H_2O + e^{-} \rightleftharpoons MH + OH^{-}$
Cathode	$Ni(OH)_2 + OH^{-} \rightleftharpoons NiO(OH) + H_2O + e^{-}$
Overall reaction	$M + Ni(OH)_2 \rightleftharpoons MH + NiO(OH)$

various applications such as consumer electronics, power tools, and EVs. While newer technologies such as LIBs have gained prominence in the EV market for their higher energy densities and lighter weight, NiMH batteries continue to find applications in certain EV models, particularly HEVs. In HEVs, NiMH batteries are often utilized in conjunction with internal combustion engines to provide additional power and improve fuel efficiency through regenerative braking, where kinetic energy is recovered during braking and stored in the batteries for later use. Despite facing competition from LIBs, NiMH batteries remain a dependable and cost-effective option for certain EV applications, highlighting their enduring significance in the evolving landscape of automotive electrification.

1.3.5 Supercapacitors

Supercapacitors represent an advanced energy storage technology that presents unique advantages compared to conventional batteries and capacitors. Supercapacitors store and deliver energy through the electrostatic double-layer capacitance and pseudo-capacitance processes, unlike batteries which store energy through chemical reactions and capacitors which store energy electrostatically. Fundamentally, supercapacitors are comprised of two electrodes, typically constructed from porous carbon materials, submerged in an electrolyte solution.

The working principle of supercapacitors involves the accumulation of charges at the interface between the electrode surface and the electrolyte. The ions from the electrolyte are attracted to the oppositely charged electrode, forming a double layer of charge, when a voltage is applied across the electrodes. This electrostatic double-layer capacitance accounts for

Table 1.10 Performance comparison of supercapacitor and lithium-ion battery.

Function	Super-caps	Li-ion battery
Charging time	1–10 seconds	10–60 min
Life-time	1 million cycles	500 or higher cycles
Voltage of cell	2.3–2.75 V	3.6 V
Specific energy	5 Wh/kg	120–240 Wh/kg
Specific power	Up to 10,000	1000–3000
Cost per kWh	$10,000	$250–$1,000
Service period	10–15 years	5–10 years
Range of charging temperature	−40 °C–65 °C	0 °C–45 °C
Range of discharging temperature	−40 °C–65 °C	−20 °C–60 °C

most of the energy storage in supercapacitors. In addition to double-layer capacitance, some supercapacitors also exhibit pseudo-capacitance, which involves reversible redox reactions at the electrode–electrolyte interface, further enhancing energy storage capacity. The chemical reactions involved in the operation of supercapacitors depend on the specific electrode materials and electrolytes used. In the case of carbon-based electrodes with an aqueous electrolyte, the charging process primarily involves the adsorption of ions onto the surface of the carbon electrodes, leading to the formation of the electric double layer. The discharge process occurs when the stored ions are released from the electrodes, generating an electric current. The overall operation of supercapacitors enables them to deliver high power densities and rapid charge–discharge cycles, making them ideal for applications requiring bursts of energy or frequent cycling. In EVs, supercapacitors find applications in hybrid energy storage systems (HESSs) alongside batteries. Supercapacitors adeptly store and discharge energy in regenerative braking scenarios, wherein kinetic energy is transformed into electrical energy and preserved for subsequent utilization. This process enhances overall energy efficiency and prolongs battery lifespan. Additionally, supercapacitors can provide supplemental power for acceleration or high-demand scenarios, enhancing the performance and longevity of EVs. The performance comparison of supercapacitor and lithium-ion battery is shown in Table 1.10. Although supercapacitors have high power density and rapid charging capabilities, they typically exhibit lower energy densities when compared to batteries. This limitation restricts their standalone application for long-range EVs. The advantages and disadvantages of automotive grade supercapacitors are listed out in Table 1.11. However, their combination with batteries in HESSs offers a synergistic solution for addressing the dynamic power requirements of modern EVs.

1.3.6 Flywheel energy storage

Flywheel energy storage systems represent an innovative approach to storing and releasing energy, offering unique advantages in various applications, including EVs. A flywheel energy storage system typically consists of a heavy rotor, often made of steel or composite materials, mounted on bearings within a vacuum or gas-filled enclosure to minimize friction and air resistance. When energy is supplied to the system, typically through an electric motor, the rotor accelerates, storing kinetic energy in its rotational motion. Conversely, when energy is needed, the rotational energy of the flywheel

Table 1.11 Advantages and disadvantages of automotive grade supercapacitors.

Advantages	Almost infinite cycle life; can be used a million times
	Higher specific power and lower resistance permit SC to feed high-load current
	Charging periods in seconds
	Simple charging draws only needed current i.e., it is not subjected to overcharge
	Safe
	Brilliant lower-temperature charging and discharging performance
Disadvantages	Lower specific energy; grips a portion of conventional battery
	Linearized discharge voltage averts using the full energy
	Higher self-discharge rate
	Lower voltage of cell requires series connection along voltage balancing
	Higher price per watt

can be converted back into electricity using a generator connected to the system [37], [38].

The working principle of flywheel energy storage revolves around the conversion of electrical energy into kinetic energy during charging and the reverse process during discharging. In the charging phase, electrical energy is used to accelerate the flywheel rotor to a high rotational speed, thereby storing energy in the form of rotational kinetic energy. During the discharging phase, the rotational energy of the spinning flywheel is converted back into electrical energy as the rotor slows down, powering an electric generator connected to the system. This process allows for the storage and retrieval of energy with high efficiency and minimal losses. The efficiency of flywheel ESSs can be influenced by factors such as bearing friction, aerodynamic drag, and material properties, all of which can impact the overall performance and energy storage capacity of the system [39].

Flywheel ESSs offer several potential applications for EVs. One key application is in regenerative braking, where kinetic energy from the vehicle's motion is captured and stored in the flywheel during braking events [40]. This stored energy can be harnessed to aid acceleration or offer supplementary power during periods of high demand, thereby diminishing dependence solely on the battery pack and enhancing overall energy efficiency. Furthermore, flywheel ESSs can function as a backup power reservoir in EVs, furnishing emergency power in instances of battery failure or depletion. Despite their potential benefits, flywheel ESSs may face challenges such as weight, size, and cost, which could limit their widespread adoption in EVs. However, ongoing advancements in materials, design, and technology are expected to continue improving the performance and viability of flywheel energy storage for various applications.

1.3.7 Thermal energy storage

Thermal energy storage systems (TESSs) offer a versatile and efficient method for capturing, storing, and utilizing heat energy, providing a range of applications across various sectors, including EVs. At its core, TESS involves the storage of thermal energy in a medium, typically a solid, liquid, or phase change material, during periods of excess heat generation or availability [41]. This energy can then be retrieved and used for heating, cooling, or power generation as needed.

TESSs offer a versatile and efficient method for capturing, storing, and utilizing heat energy, providing a range of applications across various sectors, including EVs. At its core, TESS involves the storage of thermal energy in a medium, typically a solid, liquid, or phase change material, during periods of excess heat generation or availability [41]. This stored thermal energy can then be retrieved and utilized for heating, cooling, or power generation as needed. The working principle of TESS hinges on the ability of certain materials to absorb, store, and release thermal energy through reversible phase changes or chemical reactions. Phase change materials (PCMs), for instance, undergo a transition between solid and liquid states as they absorb or release heat, enabling efficient energy storage. When heat is supplied to a PCM, such as during charging, it undergoes a phase change from solid to liquid, storing thermal energy in the process. Conversely, during discharging, the PCM releases heat as it solidifies, providing a continuous and reliable source of thermal energy [42]. The chemical reactions involved in thermal energy storage depend on the specific type of material used and the nature of the thermal energy storage process. For example, in latent heat storage systems using PCMs, the phase change reaction involves the absorption or release of heat energy without a temperature change. This can be represented by eqn (1.1):

$$\text{Solid} \rightleftharpoons \text{Liquid} + \text{Heat}. \tag{1.1}$$

The efficiency and performance of TESSs are influenced by factors such as the thermal conductivity and heat capacity of the storage medium, the insulation properties of the storage container, and the efficiency of the heat transfer mechanisms. By optimizing these parameters, TESSs can effectively capture and utilize thermal energy with minimal losses.

TESSs offer several potential applications. One key application is in cabin heating and cooling, where excess heat generated by the propulsion system of the vehicle, or external sources can be captured and stored during operation. This stored thermal energy can then be used to maintain comfortable cabin temperatures during periods of low energy demand or when the vehicle is stationary, reducing the reliance on the vehicle's primary battery pack for climate control. Additionally, TESSs can be integrated into EV charging infrastructure to provide pre-conditioning of batteries, optimizing their performance and longevity. Despite their potential benefits, challenges such as cost, weight, and space requirements may limit the widespread adoption of TESSs in EVs. However, ongoing research and development efforts aimed at improving the efficiency, reliability, and scalability of TESS technologies hold promise for their integration into future generations of EVs.

1.3.8 Graphene-based batteries

Graphene-based batteries represent a cutting-edge advancement in battery technology, leveraging the unique properties of graphene, a single layer of carbon atoms arranged in a hexagonal lattice structure. The exceptional conductivity, high surface area, and mechanical strength of graphene make it an ideal candidate for enhancing the performance and efficiency of batteries [43]. Graphene-based batteries typically consist of electrodes composed of graphene or graphene-derived materials, such as graphene oxide or graphene quantum dots, integrated into traditional battery architectures, such as lithium-ion or lithium-sulfur batteries.

The working principle of graphene-based batteries involves the efficient transport of ions and electrons within the battery, facilitated by the conductive properties of graphene. In graphene-based LIBs, during charging, lithium ions migrate from the positive electrode (cathode) to the negative electrode (anode), where they are intercalated into the graphene structure. This process involves the transfer of electrons, which flow through the external circuit to power electrical devices or vehicles. Conversely, during discharging, lithium ions are released from the graphene structure at the anode and migrate back to the cathode, while electrons flow in the opposite direction, generating an electric current [44]. The chemical reactions involved in graphene-based batteries depend on the specific battery chemistry and electrode materials used. In LIBs with graphene-based anodes, for instance, the charging process involves the intercalation of lithium ions into the graphene lattice, which can be represented by eqn (1.2) given below:

$$Li^{+} + e^{-} + Graphene \rightleftharpoons Li - Graphene. \tag{1.2}$$

The reversible nature of these reactions allows graphene-based batteries to store and release energy efficiently over multiple charge–discharge cycles, resulting in enhanced battery performance and longevity [45]. Graphene-based batteries offer numerous potential applications in EVs. By incorporating graphene-based materials into battery electrodes, EV manufacturers can achieve higher energy densities, enabling longer driving ranges on a single charge. Additionally, the high conductivity and rapid charge transfer properties of graphene can facilitate faster charging times, reducing the downtime associated with recharging EV batteries. Furthermore, graphene-based batteries have the potential to address challenges such as battery degradation and capacity fade, leading to improved reliability and lifespan of EV powertrains. Overall, the integration of graphene-based batteries into EVs holds promise for advancing the electrification of transportation and accelerating the transition to a sustainable energy future.

1.3.9 Flow batteries

Flow batteries represent a sophisticated class of rechargeable ESSs that offer unique advantages in terms of scalability, flexibility, and durability [11]. Unlike conventional batteries where energy storage occurs within enclosed cells, flow batteries store energy in external tanks of liquid electrolytes, which are circulated through electrochemical cells during charging and discharging cycles. Flow batteries typically consist of two electrolyte solutions, each housed in separate tanks, and an electrochemical cell comprised of a cathode and an anode separated by an ion-selective membrane.

The working principle of flow batteries revolves around the redox reactions that occur between the electrolyte solutions and the electrodes within the electrochemical cell. During charging, the electrolyte solution undergoes oxidation at the anode which releases electrons, while the other solution undergoes reduction at the cathode which accepts electrons. This process results in the transfer of ions between the electrolytes and the electrodes, storing energy in the form of chemical potential energy. During discharging, the stored energy is released as the reverse redox reactions occur which generates electric current that can be utilized to power electrical devices or vehicles [46]. The chemical reactions involved in flow batteries depend on the specific redox couples used in the electrolyte solutions. Common redox

Table 1.12 Electrochemical reactions of flow battery.

Electrode	Electrochemical reactions
Anolyte	$A^{n+} + xe^{-} \rightleftharpoons A^{n-x}$
Catholyte	$B^{m+} - xe^{-} \rightleftharpoons B^{m+x}$
Overall reaction	$B^{m+} + A^{n+} \rightleftharpoons B^{m+x} + A^{n-x}$

couples used in flow batteries include all-vanadium, iron-chromium, zinc-bromine, hydrogen-bromine, and organic molecules [47]. For example, in a vanadium redox flow battery (VRFB), the charging and discharging reactions involve the interconversion between vanadium ions in different oxidation states. Table 1.12 illustrates the typical reactions that occur in the anolyte and catholyte in flow batteries, with A and B denoting the active redox species in the anode and cathode, respectively [47].

Flow batteries offer several potential advantages such as higher energy density, fast charging capabilities, and long cycle life. By integrating flow battery systems into EVs, manufacturers can potentially overcome the limitations of traditional LIBs, such as limited driving range and long charging times. Flow batteries also have immense potential for supporting fast-charging applications [48].

1.3.10 Aluminum-ion batteries

Aluminum-ion batteries (AIBs) are a novel class of rechargeable energy storage devices that have gained significant attention in recent years due to their potential as an alternative to traditional LIBs. AIBs utilize aluminum ions instead of lithium ions as the charge carriers, offering several advantages such as abundant resource availability, low cost, and reduced safety concerns associated with lithium-based chemistries [49]. These batteries typically consist of a cathode material capable of intercalating aluminum ions, an anode material such as aluminum metal or graphite, and an electrolyte solution containing an aluminum salt.

The working principle of aluminum-ion batteries involves the reversible insertion and extraction of aluminum ions into and from the cathode material during charging and discharging cycles, respectively. During charging, aluminum ions are extracted from the anode and transported through the electrolyte to the cathode, where they are intercalated into the cathode material, typically a metal oxide or sulfide. This process is accompanied by the flow of electrons through the external circuit, storing energy in the battery. During discharging, aluminum ions are released from the cathode and migrate back to the anode, where they are re-intercalated into the anode material, while electrons flow in the opposite direction, generating electric current that can be utilized to power electrical devices or vehicles [50].

Aluminum-ion batteries offer several potential advantages over LIBs, including lower cost, improved safety, and higher energy density. By utilizing aluminum as the charge carrier instead of lithium, AIBs can potentially reduce the dependence on scarce lithium resources and alleviate concerns related to lithium supply chain constraints. Additionally, aluminum-ion batteries have the potential to offer higher energy densities than LIBs. Furthermore, AIBs may find applications in fast-charging infrastructure for EVs, enabling rapid refueling and reducing charging times for drivers [51], [52].

1.3.11 Zinc-air batteries

Zinc-air batteries represent a promising class of energy storage devices that have garnered significant attention for their high energy density, low cost, and environmental friendliness [53]. These batteries operate on the principle of electrochemical reactions involving the oxidation of zinc metal and the reduction of oxygen from the air. Zinc-air batteries typically consist of a zinc anode, a porous carbon-based cathode, and an alkaline electrolyte solution. The working principle of zinc-air batteries involves the electrochemical conversion of chemical energy into electrical energy through redox reactions occurring at the anode and cathode during discharge.

Zinc-air batteries are particularly attractive for large-scale energy storage applications, such as grid-level energy storage and backup power systems. However, zinc-air batteries face challenges related to limited cycle life, performance degradation over time, corrosion passivation, and issues with zinc dendrite formation during charging, which can affect battery efficiency and longevity [54]. Despite these challenges, ongoing research and development efforts aimed at improving electrode materials, electrolyte formulations, and battery designs are expected to drive the commercialization and deployment of zinc-air batteries in various applications, including EVs.

1.3.12 Sodium-ion batteries

Sodium-ion batteries (SIBs) have emerged as a promising alternative to LIBs due to their abundance, low cost, and similar electrochemical properties [55]. SIBs operate on the same principles as LIBs, utilizing the intercalation and deintercalation of ions within the battery electrodes to store and release electrical energy. Typically, SIBs consist of a sodium-containing cathode material, such as sodium cobalt oxide or sodium iron phosphate, an anode material capable of hosting sodium ions, such as hard carbon or sodium titanate, and an electrolyte solution containing sodium salt. During charging, sodium ions are extracted from the cathode and intercalated into the anode, while electrons flow through the external circuit, storing energy in the battery. Conversely, during discharging, the stored sodium ions are released from the anode and migrate back to the cathode, generating an electric current that can be utilized to power electrical devices or vehicles [56].

SIBs offer several potential advantages over LIBs, including lower cost and reduced reliance on scarce lithium resources. By utilizing sodium as the charge carrier instead of lithium, SIBs can potentially address concerns related to lithium supply chain constraints and mitigate issues associated with fluctuating lithium prices. Additionally, SIBs have the potential to offer higher energy densities than LIBs, enabling longer driving ranges on a single charge for EVs. Despite their promising attributes, challenges such as electrode material optimization, electrolyte stability, and cycle life durability may currently limit the widespread adoption of SIBs in EVs [14].

1.3.13 Molten salt batteries

Molten salt batteries represent an innovative class of ESSs that operate at high temperatures, utilizing molten salts as both the electrolyte and the active material. These batteries typically consist of two electrodes immersed in a molten salt electrolyte, which is maintained at elevated temperatures to enable ion mobility and electrochemical reactions. The working principle of molten salt batteries involves the reversible electrochemical deposition and dissolution of metal ions at the electrodes during charging and discharging cycles [15].

During charging, metal ions from the molten salt electrolyte are electrochemically deposited onto one of the electrodes, while electrons flow through the external circuit, storing energy in the battery. Conversely, during discharging, the stored metal ions are dissolved from the electrode back into the molten salt electrolyte, releasing electrons that flow through the external circuit to power electrical devices or vehicles. The overall reaction can be represented by the following equation (eqn (1.3)):

$$M^{z+} + ze^- \rightleftharpoons M, \tag{1.3}$$

where M represents the metal ion, z is the charge of the metal ion, and e^- represents an electron.

Molten salt batteries offer several advantages over conventional battery technologies, including high energy density, long cycle life [57]. Additionally, molten salt batteries are inherently safe and resistant to thermal runaway due to the high operating temperatures required for their operation. However, they also face challenges related to high operating temperatures, such as thermal management and material compatibility issues.

1.3.14 Compressed air energy storage

Compressed air energy storage (CAES) is a promising technology for storing large amounts of energy in the form of compressed air, offering a viable solution for grid-scale energy storage and renewable energy integration. CAES systems typically consist of two main components: a compressor to compress air and store it in underground caverns or above-ground tanks, and a turbine-generator unit to expand the stored compressed air and generate electricity when needed. During periods of low electricity demand or excess renewable energy generation, surplus electricity is used to power the compressor, which compresses ambient air and stores it under high pressure [58].

The working principle of CAES involves the conversion of electrical energy into potential energy stored in the form of compressed air during charging, and the conversion of this stored energy back into electrical energy during discharging. When electricity demand increases or renewable energy generation decreases, the stored compressed air is released from the storage vessel and expanded through the turbine, driving the generator to produce electricity [59]. This process can be represented by the following equations (eqn (1.4) and (1.5)) of compression and expansion:

$$\text{Compression} : \text{Electricity} + \text{Air} \rightarrow \text{CompressedAir}. \tag{1.4}$$

$$\text{Expansion} : \text{CompressedAir} \rightarrow \text{Electricity} + \text{ExhaustAir}. \tag{1.5}$$

CAES systems offer several advantages such as, long-duration storage capabilities, high efficiency, and low environmental impact. Additionally, CAES systems can be deployed at scale and integrated into existing infrastructure, making them suitable for large-scale renewable energy integration and grid stability applications [60]. However, CAES systems also face challenges related to energy losses during compression and expansion, as well as site-specific constraints for underground storage caverns.

CAES technology has been explored primarily for stationary energy storage applications rather than onboard vehicle use. However, CAES systems could potentially be integrated into EV charging infrastructure to provide fast-charging capabilities and grid support services. By storing compressed air during off-peak hours and releasing it during peak demand periods, CAES systems could help mitigate grid congestion and provide supplemental power for EV charging stations. Furthermore, CAES technology could enable the development of mobile energy storage solutions for EV charging at remote locations or temporary events. Despite these potential applications, further research and development are needed to optimize CAES technology for EV charging and address challenges such as energy efficiency, cost, and infrastructure requirements [61].

1.3.15 Thermal electric generators

Thermoelectric generators (TEGs) are solid-state devices that convert heat energy directly into electrical energy through the Seebeck effect, offering a promising solution for waste heat recovery and power generation in various applications. TEGs typically consist of thermoelectric materials arranged in a series of p-type and n-type semiconductors connected electrically in a circuit. When there is a temperature gradient across the thermoelectric material, a voltage potential is generated, driving electrons to flow from the hot side to the cold side, thus creating an electric current. This process occurs due to the Seebeck effect, which describes the phenomenon where a temperature difference across a thermoelectric material generates a voltage proportional to the temperature gradient [62],[63].

TEGs offer several potential applications for waste heat recovery and power generation [17]. EVs typically generate excess heat during operation, particularly from the engine or powertrain components, which can be harnessed by TEGs to generate electricity and improve overall energy efficiency. By integrating TEGs into the vehicle's exhaust system or thermal management system, waste heat can be converted into useful electrical energy, reducing the vehicle's reliance on the primary battery pack and extending its driving range. Additionally, TEGs can be used to power auxiliary systems in EVs, such as heating, ventilation, and air conditioning (HVAC) systems, further enhancing energy efficiency and reducing the vehicle's overall carbon footprint. Despite their potential benefits, challenges such as low efficiency and high cost currently limit the widespread adoption of TEGs in EVs. However, ongoing research and development efforts aimed at improving thermoelectric materials and device designs are expected to drive advancements in TEG technology and accelerate their integration into future generations of EVs.

1.4 Hybrid Energy Storage System (HESS)

Hybrid energy storage system (HESS) refers to a combination of two or more ESSs used in an EV with an intelligent control system to control the power flow from each energy storage system. With the increasing requirement of power and energy requirement to provide similar or even better experience like conventional gasoline cars the use of HESS become a necessity in EVs. Primary reasons of using HESS in EVs include higher range, power, longer battery life, safe, and efficient operation. Often, it is referred that HESS also used for smooth acceleration even better than convention gasoline cars. HESS also increases the battery life by minimizing stress on battery, providing a cost optimal solution for the EVs. Typically, an advanced control system is used for controlling the flow of power. Practically all pure electric vehicles (PEVs) are dominated by galvanic electrochemical batteries typically have much lower vitality thickness (J/m^3 or J/kg) with respect to conventional gasoline vehicles [64]. Therefore, in terms of range anxiety and the reliability, gasoline cars still outperform the EVs. Here, it is expected that HESS can provide the solution to range anxiety and reliability issues in EVs [64]. Therefore, in terms of range anxiety and the reliability, gasoline cars still outperform the EVs. Here, it is expected that HESS can provide the solution to range anxiety and reliability issues in EVs.

1.4.1 Battery and supercapacitor HESS

Supercapacitors, known for their high-power density ranging from 5 to 10 kW/kg, offer the capability to deliver substantial current demands rapidly and with high efficiency, particularly suitable for applications like acceleration and regenerative braking. Their use can significantly reduce charging times. However, supercapacitors lack the energy storage capacity to serve as a primary power source in EVs. Batteries, on the other hand, can store bulk energy but may not provide instant power comparable to supercapacitors without oversizing system components. Integrating both supercapacitors and batteries into a HESS can provide higher current during acceleration or can be charged with higher current during regenerative braking while at the same time ensuring a sufficiently long running span for the vehicle.

The high specific power of supercapacitors helps alleviate the strain on battery packages and potentially enhances acceleration and performance on steep inclines. Additionally, supercapacitors assist batteries in capturing energy during regenerative braking. In various industrial applications, including EVs, challenges such as energy efficiency, performance, structural sizing, and cost-effectiveness of ESSs are paramount. For instance, in the case of the Tesla Roadster model, which weighs 2690 pounds, the battery pack alone weighs more than 900 pounds. Integrating an ultracapacitor bank can offset this weight imbalance and provide greater flexibility in distributing size and weight among different energy storage components during the design phase, allowing for the realization of anticipated storage and peak current values. Research suggests that a hybrid energy storage system consisting of two 18650-cell LIBs and two 100 F ultracapacitors can achieve a peak power of 132 W, representing a seven-fold improvement compared to lithium-battery cells alone. Ongoing research focuses on optimizing the topology of HESSs, with variations in cost, complexity, and flexibility. Generally, these systems can be classified into passive and active HESS, each with further subtypes of topology.

1.4.2 Topologies of battery and superconductor HESS

In EVs, the primary energy source remains batteries due to their high energy density, despite limitations in delivering large power pulses economically and efficiently. Batteries lack the rapid charge/discharge capabilities of supercapacitors and suffer from long charging times and shorter lifespans, especially under high charging rates which can lead to safety concerns like overheating [65]. To address these issues, battery packs in EVs are oversized and equipped with active cooling technology. However, current energy accumulation systems are designed for specific operating points, leading to compromises between various requirements. HESS have been developed to combine different energy storage technologies, typically high-energy (HE) and high-power (HP) elements to improve specific power and energy characteristics. HP storage, often employing electrical double-layer capacitors (EDLCs), facilitates rapid power fluctuations, while HE storage, typically using lithium-ion cells, ensures long-term energy supply. Hybrid energy storage systems (HESSs) can be configured using various connection topologies, including passive hybrid energy storage system (P-HESS) and active hybrid energy storage system (A-HESS) topologies, which optimize the integration of different storage units with diverse characteristics [66].

1.4.2.1 Passive HESS

A passive hybrid energy storage system (P-HESS) features a straightforward connection between a battery and a supercapacitor in parallel configuration. In this setup, the battery is directly linked to the supercapacitor and then to the traction motor. This direct connection maintains relatively constant terminal voltages at both ends, minimizing the risk of battery damage when immediate power is needed from the traction motor. However, to ensure safe operation, a DC–DC converter is inserted between the energy storage bank and the DC–AC inverter [67]. This configuration represents the simplest form of hybridization, achieved by connecting high-power (HP) and high-energy (HE) elements in parallel. The voltage levels of the battery and supercapacitor must match the rated voltage of the traction motor to optimize performance. This passive coupling of storage elements is illustrated in Figure 1.2, without the need for an intermediary converter. In automotive applications, such as the powertrain, ensuring compatibility between the operating voltage levels of different storage technologies is crucial for efficient operation [65], [68].

In a hybridized configuration combining lithium-ion battery (LIBs) with electrical double-layer capacitors (EDLCs), the working lifespan surpasses that of a battery-only setup, particularly under high dynamic discharge profiles. The voltage drop in a hybrid system is lower compared to a pure battery setup due to the inherently lower internal resistance of EDLCs [69]. In passive hybridization, where EDLCs handle peak currents during acceleration and deceleration, they act as low-pass filters, smoothing rapid voltage changes. Increasing the EDLC capacity enhances filtering properties and experimental evidence suggests it improves power quality and source efficiency. However, the usable capacity of EDLCs is limited due to direct coupling, with the operating voltage window constrained by the battery or powertrain. Consequently, achieving high utilization of stored energy, especially in low temperatures, requires an active power distribution strategy. Implementing active power

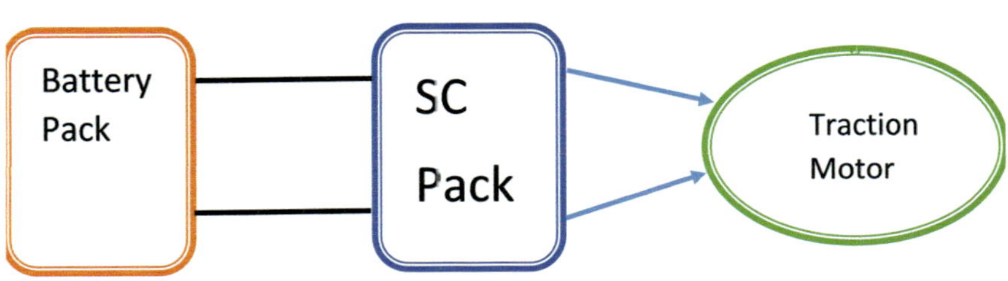

Figure 1.2 Basic passive configuration.

distribution is crucial for enhancing the performance of the energy storage system, allowing for more efficient utilization of stored energy content [70].

1.4.2.2 Active HESS

In contrast to passive hybrid energy storage system (P-HESS) topology, an active hybrid energy storage system (A-HESS) topology requires more advanced power electronic converters and controllers for batteries and supercapacitors. By employing specific methodologies or algorithms, the regulator can be adjusted to enhance the flexibility of the HESS. Unlike the direct connection of the battery and supercapacitor in P-HASS, an intermediate DC–DC converter is inserted in the A-HESS, enabling the supercapacitor to operate within a wider voltage range. However, the battery voltage remains fixed due to its direct connection with the DC–AC inverter, posing similar concerns as in P-HESS regarding the battery's protection. To address this issue, exchanging the roles of the supercapacitor and battery pack in the topology could potentially alleviate the problem. The DC–DC converter can help regulate the power flow to keep the battery within a safer operating range while the supercapacitor serves as an energy buffer. However, the operational range of supercapacitors is constrained due to their direct connection with the DC–AC inverter. Overall, the cost and performance characteristics of energy storage devices vary widely depending on the technology used, necessitating appropriate isolation to optimize their functionality, where energy storage units are potentially separated, such as through a DC–DC inverter [71].

1.4.2.3 Semi-active HESS

A semi-active hybrid energy storage system (SA-HESS) topology typically involves incorporating at least two different energy storage devices, with a portion of the energy storage device being isolated. This configuration, known as a power-isolated design, utilizes a DC–DC converter for isolation. While adding a converter to the system increases costs and requires additional installation space, this topology class offers numerous advantages. Isolated energy storage can operate optimally based on its charge and discharge characteristics. Researchers have investigated the integration of various energy storage devices, such as batteries and electrical double-layer capacitors (EDLCs) [71]. Depending on the specific energy storage configuration, subtopologies like the semi-active hybrid energy storage system (SA-HESS), parallel semi-active hybrid energy storage system (PSA-HESS) topology, battery semi-active hybrid energy storage system (BSA-HESS) topology, and the capacitor semi-active hybrid energy storage system (CSA-HESS)

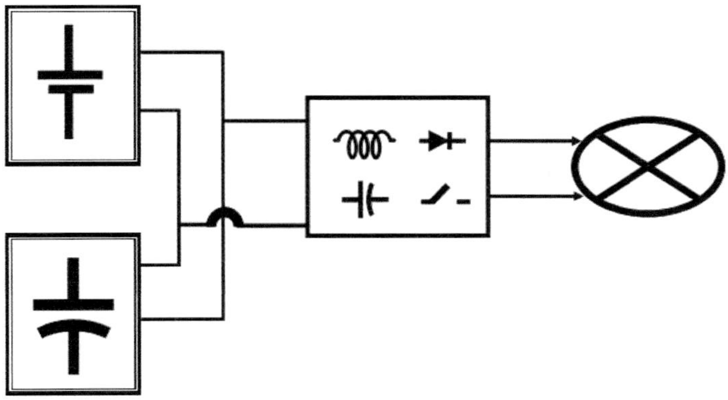

Figure 1.3 Parallel SA-HESS.

topology have been examined. These sub-topologies will be discussed in detail in the following sections [72].

1.4.2.4 Parallel semi-active HESS

In parallel semi-active hybrid energy storage system (PSA-HESS) topology, the isolation between the energy storage device and the load is achieved using a DC–DC converter, while the power storage devices are directly connected in parallel. This separation ensures that the operating voltage range of the power system is independent of the load voltage range. Consequently, this eliminates mutual dependencies and simplifies decision-making regarding energy devices and load choices. DC–DC converters are designed to accommodate the maximum power and maximum load current required by the load. When the energy device is directly connected, the voltage of both power supplies will be identical. The operational voltage range is typically determined by and restricts the usable capacity of the electrical double-layer capacitor (EDLC). Like the P-HESS configuration, the actual distribution is influenced by the internal resistance and equilibrium voltage of the two energy units. Thus, the outcomes of directly connecting the energy device are equivalent to those of P-HESS. Figure 1.3 illustrates the arrangement of system components in a parallel semi-active hybrid energy storage system.

1.4.2.5 Battery semi-active HESS

The battery semi-active hybrid energy storage system (BSA-HESS) topology removes battery storage from both the electrical double-layer capacitor

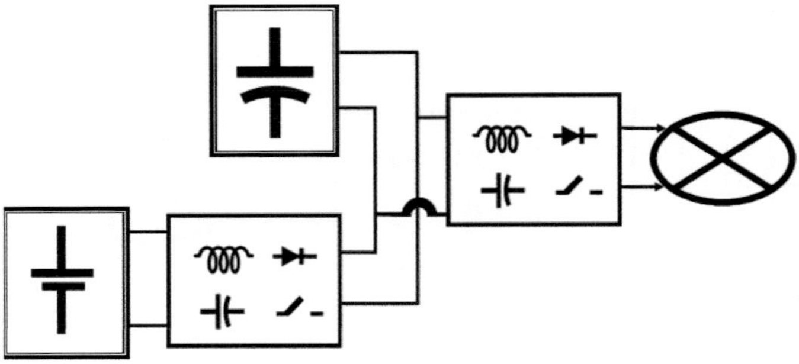

Figure 1.4 Battery SA-HESS.

(EDLC) and the load, resulting in a temporary decrease in battery charge following a lighting change. Power transfer between the battery and the load can be regulated using a DC–DC converter. Regardless of its operating capacity, the battery remains available, allowing for increased EV size. Employing appropriate maintenance techniques, the EDLC can swiftly provide short-term power output to expedite acceleration, particularly beneficial for Li-ion battery packs with nonlinear voltage profiles. Direct integration of the EDLC onto the powertrain enables adaptation to varying load voltages during acceleration and steady motion. The DC–DC converter is designed to handle significant fluctuations in load voltages. In the event of DC–DC converter failure, the load is supported by energy stored in the EDLC. However, due to the EDLC's limited capacity compared to the battery, its role as an emergency backup in a BEV may be somewhat impractical. The battery semi-active hybrid energy storage system is shown in Figure 1.4.

1.4.2.6 Capacitor semi-active HESS

Evaluating the BSA-HESS topology against the capacitor semi-active hybrid energy storage system (CSA-HESS) topology involves a shift in the arrangement of the battery and electrical double-layer capacitor (EDLC) as shown in Figure 1.5. In the CSA-HESS topology, the EDLC is separated from both the load and battery storage through a DC–DC converter, as illustrated. This converter regulates power flow and provides the EDLC with a broad operating voltage range. Unlike passive composite or direct coupling structures, the full power potential of the EDLC can be utilized in this setup, allowing it to handle maximum loads. However, this configuration necessitates DC–DC converters capable of handling maximum power and voltage levels, leading

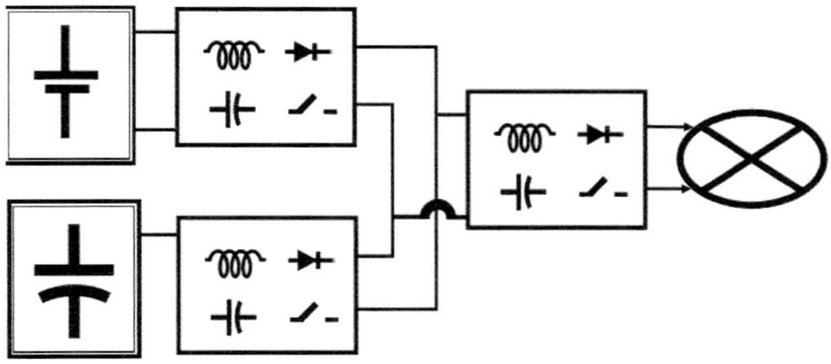

Figure 1.5 Capacitor SA-HESS.

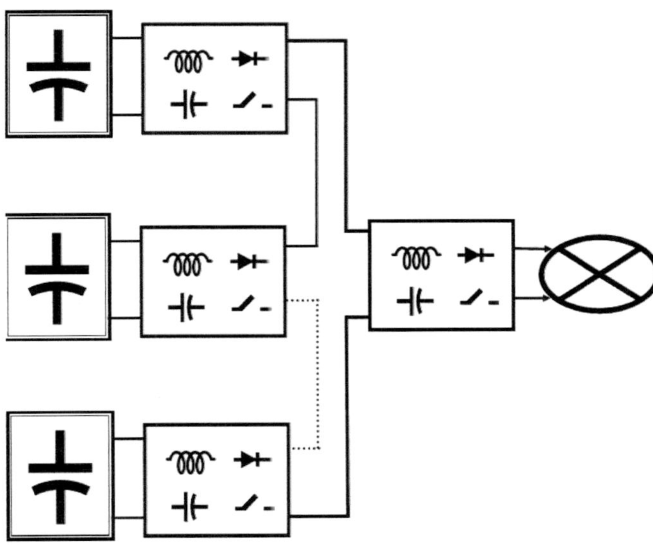

Figure 1.6 Full-active HESS.

to increased system costs. Unlike in BSA-HESS, directly combining battery storage with power in the CSA-HESS topology results in a substantially lower reduction in load voltage.

1.4.2.7 Full-active HESS

The full-active hybrid energy storage system (FA-HESS) comprises two or more diverse energy storage devices, with each storage unit separated by power electronic components as shown in Figure 1.6. This configuration is also referred to as a fully integrated framework. Typically, separation is achieved using bidirectional DC–DC converters. Compared to a SA-HESS topology, this setup incurs higher system losses due to the increased number of DC–DC converters, leading to greater complexity and costs in terms of installation space and electronics. Despite the challenges associated with expanding the number of DC–DC converters, FA-HESS offers several advantages. When operated correctly, this high-control system ensures optimal performance of all energy storage modules based on their charging and output characteristics. This precise power distribution enhances accumulation efficiency and contributes to prolonging battery life.

1.4.2.8 Modular multilevel full-active HESS

Modular multilevel full-active hybrid energy storage system (MMFA-HESS) [72] is a variant of FA-HESS topology where multiple energy storage modules are connected in series. Each module consists of energy storage devices and associated electrical components like DC–DC converters, serving as charge/discharge units. Integrated DC–DC converters enable individual modules to operate optimally according to their specific charging and discharging requirements. The incorporation of power electronics into each module allows for quick adaptation to different voltage and power levels. This topology accommodates variations in capacity, cell chemistry, internal resistance, self-discharge rate, health status, and aging behavior of each energy storage device, making it suitable for second-life applications. However, ensuring proper control of maximum train current and voltage levels at each DC–DC switching device is crucial to prevent overheating and system failures, which can lead to increased implementation costs.

1.5 Conclusion

The rapid electrification of transportation has necessitated the development of efficient and reliable ESSs for EVs. This chapter explores the evolution and advancements in various energy storage technologies tailored for EV applications.

Lithium-ion batteries have emerged as the predominant energy storage solution for EVs due to their high energy density, long cycle life, and relatively low self-discharge rate. Starting with the fundamental principles of Li-ion batteries, including their composition, electrochemical processes, and performance characteristics are discussed in this chapter. Additionally, it discusses recent innovations such as solid-state electrolytes and silicon-based anodes aimed at enhancing energy density and safety while reducing costs.

Supercapacitors, also known as ultracapacitors or electrochemical capacitors, complement Li-ion batteries by providing high power density and rapid charge/discharge capabilities. The operating principles of supercapacitors, emphasizing their ability to efficiently capture and release energy during acceleration and regenerative braking in EVs are discussed in this chapter. Furthermore, it highlights recent advancements in electrode materials, electrolytes, and device architectures to improve energy density, cycling stability, and temperature tolerance.

Hybrid energy storage systems (HESS) integrate multiple energy storage technologies, such as Li-ion batteries and supercapacitors, to synergistically leverage their respective strengths and mitigate weaknesses. The design considerations, control strategies, and performance benefits of HESS in enhancing the overall efficiency, reliability, and longevity of EV powertrains are discussed in Section 1.4 of the chapter. Moreover, it discusses emerging research on hybridization with alternative storage elements like metal-air batteries and hydrogen fuel cells to address specific application requirements and operational challenges.

Beyond conventional Li-ion batteries and supercapacitors, some emerging advanced ESSs are discussed and tailored to next-generation EVs. Examples include solid-state batteries, flow batteries, and thermal energy storage solutions, each offering unique advantages such as higher energy density, faster-charging rates, and improved safety profiles. The chapter evaluates the current state-of-the-art, ongoing research trends, and potential integration pathways for AESS into future EV platforms. In conclusion, the electrification of transportation hinges on the continued innovation and optimization of ESSs tailored for EVs. LIBs remain the cornerstone of EV powertrains, while supercapacitors and HESSs complement their performance and reliability. Looking ahead, advancements in advanced ESSs hold promise for further enhancing the range, efficiency, and sustainability of EVs, thereby accelerating the transition toward a cleaner and greener mobility landscape.

References

[1] S. Alahakoon, M. Leksell, and S. Östlund, "Emerging energy storage solutions for transportation electrification - A review," *Int. J. Eng. Technol. Innov.*, 2019.

[2] A. Masias, "Lithium-Ion Battery Design for Transportation," in *Green Energy and Technology*, 2018. doi: 10.1007/978-3-319-69950-9_1.

[3] A. Manthiram, "A reflection on lithium-ion battery cathode chemistry," *Nature Communications*. 2020. doi: 10.1038/s41467-020-15355-0.

[4] J. Tarabay and N. Karami, "Nickel Metal Hydride battery: Structure, chemical reaction, and circuit model," in *2015 3rd International Conference on Technological Advances in Electrical, Electronics and Computer Engineering, TAEECE 2015*, 2015. doi: 10.1109/TAEECE.2015.7113594.

[5] R. Mori, "Cathode materials for lithium-sulfur battery: a review," *Journal of Solid State Electrochemistry*. 2023. doi: 10.1007/s10008-023-05387-z.

[6] B. Cook, "Introduction to fuel cells and hydrogen technology," *Eng. Sci. Educ. J.*, 2002, doi: 10.1049/esej:20020601.

[7] S. Rajagopal, R. Pulapparambil Vallikkattil, M. Mohamed Ibrahim, and D. G. Velev, "Electrode Materials for Supercapacitors in Hybrid Electric Vehicles: Challenges and Current Progress," *Condensed Matter*. 2022. doi: 10.3390/condmat7010006.

[8] S. M. Mousavi G, F. Faraji, A. Majazi, and K. Al-Haddad, "A comprehensive review of Flywheel Energy Storage System technology," *Renewable and Sustainable Energy Reviews*. 2017. doi: 10.1016/j.rser.2016.09.060.

[9] G. Alva, Y. Lin, and G. Fang, "An overview of thermal energy storage systems," *Energy*. 2018. doi: 10.1016/j.energy.2017.12.037.

[10] C. Dai, G. Sun, L. Hu, Y. Xiao, Z. Zhang, and L. Qu, "Recent progress in graphene-based electrodes for flexible batteries," *InfoMat*, 2020, doi: 10.1002/inf2.12039.

[11] A. Z. Weber, M. M. Mench, J. P. Meyers, P. N. Ross, J. T. Gostick, and Q. Liu, "Redox flow batteries: A review," *J. Appl. Electrochem.*, vol. 41, no. 10, pp. 1137–1164, 2011, doi: 10.1007/s10800-011-0348-2.

[12] Y. Zhang, S. Liu, Y. Ji, J. Ma, and H. Yu, "Emerging Nonaqueous Aluminum-Ion Batteries: Challenges, Status, and Perspectives," *Advanced Materials*. 2018. doi: 10.1002/adma.201706310.

[13] C. Xia et al., "Recent Advances on Electrospun Nanomaterials for Zinc–Air Batteries," *Small Science*. 2021. doi: 10.1002/smsc.202100010.

[14] L. Zhao et al., "Engineering of Sodium-Ion Batteries: Opportunities and Challenges," *Engineering*, vol. 24, pp. 172–183, 2023, doi: 10.1016/j.eng.2021.08.032.

[15] H. Liu, X. Zhang, S. He, D. He, Y. Shang, and H. Yu, "Molten salts for rechargeable batteries," *Mater. Today*, vol. 60, pp. 128–177, 2022, doi: 10.1016/j.mattod.2022.09.005.

[16] H. Lund and G. Salgi, "The role of compressed air energy storage (CAES) in future sustainable energy systems," *Energy Convers. Manag.*, 2009, doi: 10.1016/j.enconman.2009.01.032.

[17] D. Champier, "Thermoelectric generators: A review of applications," *Energy Convers. Manag.*, vol. 140, pp. 167–181, 2017, doi: 10.1016/j.enconman.2017.02.070.

[18] H. Yang, C. N. Savory, B. J. Morgan, D. O. Scanlon, J. M. Skelton, and A. Walsh, "Chemical Trends in the Lattice Thermal Conductivity of Li(Ni, Mn, Co)O2(NMC) Battery Cathodes," *Chem. Mater.*, 2020, doi: 10.1021/acs.chemmater.0c02908.

[19] M. K. Tran, A. Dacosta, A. Mevawalla, S. Panchal, and M. Fowler, "Comparative Study of Equivalent Circuit Models Performance in Four Common Lithium-Ion Batteries: LFP, NMC, LMO, NCA," *Batter. 2021*, Vol. 7, Page 51, vol. 7, no. 3:51, Jul. 2021, doi: 10.3390/BATTERIES7030051.

[20] X. Liu, K. Li, and X. Li, "The electrochemical performance and applications of several popular lithium-ion batteries for electric vehicles - A review," in *Communications in Computer and Information Science: Advances in Green Energy Systems and Smart Grid*, Springer Singapore, 2018, pp. 201–213. doi: 10.1007/978-981-13-2381-2_19.

[21] A. Zülke et al., "High-Energy Nickel-Cobalt-Aluminium Oxide (NCA) Cells on Idle: Anode- versus Cathode-Driven Side Reactions," *Batter. Supercaps*, 2021, doi: 10.1002/batt.202100046.

[22] C. Li et al., "Optimization of LiFePO 4 cathode material based on phosphorus doped graphite network structure for lithium ion batteries," *Ionics (Kiel).*, vol. 25, no. 3, pp. 927–937, Mar. 2019, doi: 10.1007/S11581-018-2744-7/FIGURES/12.

[23] F. Forte, M. Pietrantonio, S. Pucciarmati, M. Puzone, and D. Fontana, "Lithium iron phosphate batteries recycling: An assessment of current status," *Crit. Rev. Environ. Sci. Technol.*, 2021, doi: 10.1080/10643389.2020.1776053.

[24] Y. Zhang et al., "Research Development on Spinel Lithium Manganese Oxides Cathode Materials for Lithium-Ion Batteries," *J. Electrochem. Soc.*, vol. 170, no. 9, 2023, doi: 10.1149/1945-7111/acf8fd.

[25] S. Zhou, H. Han, J. Nie, M. Armand, Z. Zhou, and X. Huang, "Improving the High-Temperature Resilience of LiMn2O4 Based Batteries: LiFNFSI an Effective Salt," *J. Electrochem. Soc.*, vol. 159, no. 8, pp. A1158–A1164, Jul. 2012, doi: 10.1149/2.026208JES/XML.

[26] M. Al-saadi, J. Olmos, A. Saez-de-ibarra, J. Van Mierlo, and M. Berecibar, "Fast Charging Impact on the Lithium-Ion Batteries' Lifetime and Cost-Effective Battery Sizing in Heavy-Duty Electric Vehicles Applications," *Energies*, vol. 15, no. 4, 2022, doi: 10.3390/en15041278.

[27] P. Kurzweil and K. Brandt, "Overview of rechargeable lithium battery systems," in *Electrochemical Power Sources: Fundamentals, Systems, and Applications Li-Battery Safety*, J. Garche and K. Brandt, Eds., Elsevier, 2019, pp. 47–82. doi: 10.1016/B978-0-444-63777-2.00003-7.

[28] H. D. Yoo, E. Markevich, G. Salitra, D. Sharon, and D. Aurbach, "On the challenge of developing advanced technologies for electrochemical energy storage and conversion," *Mater. Today*, vol. 17, no. 3, pp. 110–121, Apr. 2014, doi: 10.1016/J.MATTOD.2014.02.014.

[29] A. B. Yaroslavtsev, Y. A. Dobrovolsky, N. S. Shaglaeva, L. A. Frolova, E. V Gerasimova, and E. A. Sanginov, "Nanostructured materials for low-temperature fuel cells," *Russ. Chem. Rev.*, vol. 81, no. 3, pp. 191–220, 2012, doi: 10.1070/rc2012v081n03abeh004290.

[30] T. T. Zuo et al., "A mechanistic investigation of the Li10GeP2S12|LiNi1-x-yCoxMnyO2 interface stability in all-solid-state lithium batteries," *Nat. Commun. 2021 121*, vol. 12, no. 1, pp. 1–10, Nov. 2021, doi: 10.1038/S41467-021-26895-4.

[31] B. S. Vishnugopi et al., "Challenges and Opportunities for Fast Charging of Solid-State Lithium Metal Batteries," *ACS Energy Lett.*, vol. 6, no. 10, pp. 3734–3749, 2021, doi: 10.1021/acsenergylett.1c01352.

[32] M. Zhao, B. Q. Li, X. Q. Zhang, J. Q. Huang, and Q. Zhang, "A Perspective toward Practical Lithium-Sulfur Batteries," *ACS Cent. Sci.*, vol. 6, no. 7, pp. 1095–1104, 2020, doi: 10.1021/acscentsci.0c00449.

[33] G. Zhou, H. Chen, and Y. Cui, "Formulating energy density for designing practical lithium–sulfur batteries," *Nat. Energy*, vol. 7, no. 4, pp. 312–319, 2022, doi: 10.1038/s41560-022-01001-0.

[34] M. Wild et al., "Lithium sulfur batteries, a mechanistic review," *Energy Environ. Sci.*, vol. 8, no. 12, pp. 3477–3494, 2015, doi: 10.1039/c5ee01388g.

[35] W. H. Zhu, Y. Zhu, Z. Davis, and B. J. Tatarchuk, "Energy efficiency and capacity retention of Ni-MH batteries for storage applications," *Appl. Energy*, vol. 103, pp. 307–313, 2013, doi: 10.1016/j.apenergy.2012.12.025.

[36] P. Kurzweil and J. Garche, "Overview of batteries for future automobiles," in *Lead-Acid Batteries for Future Automobiles*, 2017, pp. 27–96. doi: 10.1016/B978-0-444-63700-0.00002-7.

[37] A. A. Gharahbagh, V. Hajihashemi, J. M. R. da Silva Tavares, M. Sadi, A. K. Singh, and A. Arabkoohsar, "Flywheel energy storage," in *Future Grid-Scale Energy Storage Solutions: Mechanical and Chemical Technologies and Principles*, 2023, pp. 507–541. doi: 10.1016/B978-0-323-90786-6.00012-1.

[38] A. G. Olabi, T. Wilberforce, M. A. Abdelkareem, and M. Ramadan, "Critical review of flywheel energy storage system," *Energies*, vol. 14, no. 8, 2021, doi: 10.3390/en14082159.

[39] K. Xu, Y. Guo, G. Lei, and J. Zhu, "A Review of Flywheel Energy Storage System Technologies," *Energies 2023*, Vol. 16, Page 6462, vol. 16, no. 18, Sep. 2023, doi: 10.3390/EN16186462.

[40] M. Hedlund, J. Lundin, J. de Santiago, J. Abrahamsson, and H. Bernhoff, "Flywheel energy storage for automotive applications," *Energies*, vol. 8, no. 10, 2015, doi: 10.3390/en81010636.

[41] I. Sarbu and C. Sebarchievici, "A comprehensive review of thermal energy storage," *Sustain.*, vol. 10, no. 1, 2018, doi: 10.3390/su10010191.

[42] K. Pielichowska and K. Pielichowski, "Phase change materials for thermal energy storage," *Prog. Mater. Sci.*, vol. 65, pp. 67–123, 2014, doi: 10.1016/j.pmatsci.2014.03.005.

[43] J. Zhu, D. Yang, Z. Yin, Q. Yan, and H. Zhang, "Graphene and graphene-based materials for energy storage applications," *Small*, vol. 10, no. 17, pp. 3480–3498, 2014, doi: 10.1002/smll.201303202.

[44] L. Lavagna, G. Meligrana, C. Gerbaldi, A. Tagliaferro, and M. Bartoli, "Graphene and lithium-based battery electrodes: A review of recent literature," *Energies*, vol. 13, no. 18, 2020, doi: 10.3390/en13184867.

[45] A. Ali, F. Liang, J. Zhu, and P. K. Shen, "The role of graphene in rechargeable lithium batteries: Synthesis, functionalisation, and perspectives," *Nano Mater. Sci.*, 2022, doi: 10.1016/j.nanoms.2022.07.004.

[46] E. Sánchez-Díez et al., "Redox flow batteries: Status and perspective towards sustainable stationary energy storage," *J. Power Sources*, vol. 481, 2021, doi: 10.1016/j.jpowsour.2020.228804.

[47] A. G. Olabi et al., "Redox Flow Batteries: Recent Development in Main Components, Emerging Technologies, Diagnostic Techniques, Large-Scale Applications, and Challenges and Barriers," *Batteries*, vol. 9, no. 8, 2023, doi: 10.3390/batteries9080409.

[48] B. Khaki and P. Das, "Multi-objective optimal charging current and flow management of Vanadium Redox Flow Batteries for fast charging and energy-efficient operation," *J. Power Sources*, vol. 506, 2021, doi: 10.1016/j.jpowsour.2021.230199.

[49] D. Yuan, J. Zhao, W. Manalastas, S. Kumar, and M. Srinivasan, "Emerging rechargeable aqueous aluminum ion battery: Status, challenges, and outlooks," *Nano Mater. Sci.*, vol. 2, no. 3, pp. 248–263, 2020, doi: 10.1016/j.nanoms.2019.11.001.

[50] T. Leisegang et al., "The aluminum-ion battery: A sustainable and seminal concept?," *Front. Chem.*, vol. 7, 2019, doi: 10.3389/fchem.2019.00268.

[51] X. Yang and C. Huang, "Progress in improving the performance of inorganic cathodes for aluminium-ion batteries," *J. Energy Storage*, vol. 78, Feb. 2024, doi: 10.1016/J.EST.2023.110069.

[52] J. S. Edge et al., "Lithium ion battery degradation: what you need to know," *Phys. Chem. Chem. Phys.*, vol. 23, no. 14, pp. 8200–8221, Apr. 2021, doi: 10.1039/D1CP00359C.

[53] M. Al-Amin, S. Islam, S. U. A. Shibly, and S. Iffat, "Comparative Review on the Aqueous Zinc-Ion Batteries (AZIBs) and Flexible Zinc-Ion Batteries (FZIBs)," *Nanomaterials*, vol. 12, no. 22, 2022, doi: 10.3390/nano12223997.

[54] X. Zhang et al., "Comprehensive review on zinc-ion battery anode: Challenges and strategies," *InfoMat*, vol. 4, no. 7, 2022, doi: 10.1002/inf2.12306.

[55] P. K. Nayak, L. Yang, W. Brehm, and P. Adelhelm, "From Lithium-Ion to Sodium-Ion Batteries: Advantages, Challenges, and Surprises," *Angew. Chemie - Int. Ed.*, vol. 57, no. 1, pp. 102–120, 2018, doi: 10.1002/anie.201703772.

[56] A. M. Skundin, T. L. Kulova, and A. B. Yaroslavtsev, "Sodium-Ion Batteries (a Review)," *Russ. J. Electrochem.*, vol. 54, 2018, doi: 10.1134/S1023193518020076.

[57] L. Wagner, "Overview of Energy Storage Technologies," in *Future Energy: Improved, Sustainable and Clean Options for our Planet*, 2013, pp. 613–631. doi: 10.1016/B978-0-08-099424-6.00027-2.

[58] L. Chen, T. Zheng, S. Mei, X. Xue, B. Liu, and Q. Lu, "Review and prospect of compressed air energy storage system," *J. Mod. Power Syst. Clean Energy*, vol. 4, pp. 529–541, 2016, doi: 10.1007/s40565-016-0240-5.

[59] E. Barbour and D. L. Pottie, "Adiabatic compressed air energy storage technology," *Joule*, vol. 5, no. 8, pp. 1914–1920, 2021, doi: 10.1016/j.joule.2021.07.009.

[60] S. Chen, H. R. Rahbari, A. Arabkoohsar, T. Zhu, and G. Chen, "Compressed air energy storage system," in *Future Grid-Scale Energy Storage*

Solutions: Mechanical and Chemical Technologies and Principles, 2023, pp. 269–308. doi: 10.1016/B978-0-323-90786-6.00004-2.

[61] F. Khalafian et al., "Capabilities of compressed air energy storage in the economic design of renewable off-grid system to supply electricity and heat costumers and smart charging-based electric vehicles," *J. Energy Storage*, vol. 78, Feb. 2024, doi: 10.1016/J.EST.2023.109888.

[62] Y. Liu, J. R. Riba, M. Moreno-Eguilaz, and J. Sanllehí, "Application of Thermoelectric Generators for Low-Temperature-Gradient Energy Harvesting," *Appl. Sci.*, vol. 13, no. 4, 2023, doi: 10.3390/app13042603.

[63] N. Jaziri, A. Boughamoura, J. Müller, B. Mezghani, F. Tounsi, and M. Ismail, "A comprehensive review of Thermoelectric Generators: Technologies and common applications," *Energy Reports*, vol. 6, pp. 264–287, 2020, doi: 10.1016/j.egyr.2019.12.011.

[64] R. Bindu, S. Patil, and S. Thale, "Design and control of power conversion system for electric vehicle application," in *Proceedings of 2017 IEEE International Conference on Technological Advancements in Power and Energy: Exploring Energy Solutions for an Intelligent Power Grid, TAP Energy 2017*, 2018. doi: 10.1109/TAPENERGY.2017.8397264.

[65] S. Bahravar, H. M. Mahery, E. Babaei, and M. Sabahi, "Mathematical modeling and transient analysis of DC–DC buck-boost converter in CCM," in *India International Conference on Power Electronics, IICPE*, 2012. doi: 10.1109/IICPE.2012.6450404.

[66] X. Xia, X. Zhao, H. Zeng, and X. Zeng, "A novel design of hybrid energy storage system for electric vehicles," *Chinese J. Electr. Eng.*, 2018, doi: 10.23919/CJEE.2018.8327370.

[67] A. Khaligh and Z. Li, "Battery, ultracapacitor, fuel cell, and hybrid energy storage systems for electric, hybrid electric, fuel cell, and plug-in hybrid electric vehicles: State of the art," *IEEE Trans. Veh. Technol.*, 2010, doi: 10.1109/TVT.2010.2047877.

[68] R. Katuri and S. Gorantla, "Performance and Comparative Analysis of Hybrid Controllers Implemented to Hybrid Energy Storage System of Electric Vehicles," in *5th International Conference on Electrical Energy Systems, ICEES 2019*, 2019. doi: 10.1109/ICEES.2019.8719308.

[69] P. G. and J. P., "Multiple Energy Sources Hybridization: The Future of Electric Vehicles?," in *New Generation of Electric Vehicles*, 2012. doi: 10.5772/53359.

[70] S. Tiwari, S. Gupta, S. Jain, and B. Singh, "Artificial neural network based series active power filter for self excited induction generator," *J. Renew. Sustain. Energy*, 2015, doi: 10.1063/1.4905699.

[71] A. Aljanad, A. Mohamed, T. Khatib, A. Ayob, and H. Shareef, "A novel charging and discharging algorithm of plug-in hybrid electric vehicles considering vehicle-to-grid and photovoltaic generation," *World Electr. Veh. J.*, 2019, doi: 10.3390/wevj10040061.

[72] F. Akar, Y. Tavlasoglu, E. Ugur, B. Vural, and I. Aksoy, "A Bidirectional Nonisolated Multi-Input DC–DC Converter for Hybrid Energy Storage Systems in Electric Vehicles," *IEEE Trans. Veh. Technol.*, 2016, doi: 10.1109/TVT.2015.2500683.

2

Electric Vehicle Battery Chargers

Roscia Jeya Shiney Jayakumar[1] and Allwyn Clarence Asis Arul[2]

[1]Department of Electronics and Communication Engineering, P.S.R. Engineering College, India
[2]Department of Electrical and Electronics Engineering, Mepco Schlenk Engineering College, India
E-mail: shineyroscia@gmail.com; iamallwyn@mepcoeng.ac.in

Abstract

An extensive study of electric vehicle (EV) battery chargers is presented in this chapter, with an emphasis on the essential function of EV battery chargers in electric mobility. It begins by summarizing the different kinds of charging solutions that are available, from high speed DC fast chargers to Level 1 and Level 2 chargers, and describing their distinctive use cases and advantages. This chapter provides insightful information about various EV charging techniques and their vital function in facilitating the shift to sustainable mobility by combining the most recent research with real world examples. The discussion focuses on recent advances in smart charging and inductive charging systems, among other technologies that increase charging speed and efficiency. The chapter addresses the topologies of electric vehicle converters, emphasizing their vital function in controlling the flow of power between the batteries, electric motor and charging systems. It begins by categorizing several converter types, such as buck, boost, and bi-directional converters, highlighting their unique applications and advantages in EV systems. The chapter also discusses the most recent advances in charging technology, including wireless charging and smart charging solution which have increased the efficiency. The chapter elaborates the challenges of accessibility, standardization and integration of renewable energy into the charging network, which are limitations to the wide use of chargers. The architectural framework of electric vehicle (EV) battery chargers is also examined in this chapter, with an emphasis on essential components and design principles. The incorporation of smart technologies such as energy management systems and grid connectivity that improve overall user experience and maximize energy utilization is also addressed in this chapter. The concept and operation of smart battery chargers are also focused in this chapter, with an emphasis on how important they are for maximizing the charging of electric vehicles and other devices. Thus this chapter provides insightful information about the evolving field of EV charging techniques and their crucial role in facilitating the shift to sustainable mobility by combining the most recent research with real world examples.

Key words: Electric vehicle, Wireless power transfer, Thermal management strategies, Battery swap station, Smart battery charger, Vehicle to grid technology, Hybrid electric vehicles.

2.1 Introduction

Electric vehicles (EVs) serve as a possible substitute for internal combustion engine (ICE) vehicles in the current scenario. Electric vehicle usage decreases the petroleum consumption and satisfies the regulations on emissions, reducing the global warming effect. There are many factors that restrict the usage of EVs, such as reduced battery capacity, high cost of the batteries, and increase in charging time, inadequate charging infrastructure, and short driving capability [1–3]. In the present scenario, automobile manufacturers are interested in marketing hybrid electric vehicles (HEVs) to overcome the drawbacks of conventional ICE vehicles. The HEVs has different configurations such as parallel, series, series-parallel, and complex [4]. The main consideration in HEVs is to enhance the driving efficiency, in which both ICE and electric motor drives take part in the working process. HEVs are not entirely emission free and fall under the classification of very low emissions cars. In a series hybrid electric vehicle, the internal combustion engine powers the electric generator while the electric motor provides the traction power supply. The energy needed to charge the batteries is generated by the internal combustion engine. The engine and electric motor of a parallel hybrid electric vehicle are connected, allowing both to run at high speeds simultaneously. According to the operating conditions, the internal combustion engine in a series-parallel configuration powers the vehicle and the motor in addition to producing electricity that is stored in the battery.

Based on the hybridization factor, HEVs can be divided into four categories: micro, mild, energy, and hybrid grid. The ratio of the peak electrical power of the vehicle to the overall electrical and mechanical power is known as the hybridization factor [5]. Power hybrids have a larger hybridization factor than mild hybrids, which ranges from 10% to 25%. Micro hybrids have an overall hybridization factor of between 5% and 10%. Since HEVs require grid power to charge, they are also known as plug-in hybrid electric cars, or PHEVs. Grid electricity can be used by HEVs to replenish their energy storage system. PHEVs have a high-density energy storage system that is capable of being charged externally and run only on electricity, extending their lifespan and lowering the fuel economy.

2.2 Electric Vehicle (EV)

Electric vehicles (EVs) can determine their state of charge (SOC) in accordance with road inclination, the weight of the vehicle, and the resistance offered by the wind.

In comparison to HEVs, the entire EV is powered by a battery pack, which helps to mitigate the impact of global warming. In an electric vehicle, the kinetic energy of the vehicle is converted into electrical energy by the regenerative braking system and stored in the battery. The motor, which propels the car forward, is run by the battery.

The electric vehicles are much more suitable for urban driving since, when the vehicle often starts and stops, it recaptures some of the kinetic energy into the battery [6]. There are two modes of operation for an electric vehicle. In all time modes, every time the vehicle is in motion, the battery propels it. In the deceleration mode, the vehicle uses regenerative technology to bring back the kinetic energy into the battery.

Though the fuel cell tank's energy source is hydrogen, the fuel cell electric vehicle (FCEV) is identical to an EV in every way. Because they don't produce any pollutants, FCEVs are regarded as zero-emission vehicles [7]. Based on power train arrangement, fuel cell electric vehicles (FCEVs) and fuel cell hybrid electric vehicles (FCHEVs) are the two categories of FCEV.

The charging and discharging of ESS is dependent of the power demand of the vehicle and supply. FCHEV is suitable for high-speed operation when compared with FCEV, which is appropriate for smooth power demand applications and low-speed applications. The disadvantage of the FCHEV is the extra weight added to the system owing to the energy storage system in the power train [8].

2.2.1 Battery charging modes for EVs

The Society of Automotive Engineers (SAE) and the International Electrotechnical Commission (IEC) are in charge of standardizing EV connections to power networks. Among the most crucial requirements are:

- IEC 61851-1 [9]: This standard describes the vehicle connection, device working circumstances, and supply charging level.
- IEC 62196-1 [10]: This standard specifies vehicle connections, plugs, socket-outlets, and inlets for plug-in electric vehicles and plug-in hybrid electric vehicles.
- IEC 61980-1 [11]: This standard can be used to operate equipment that transfers electricity wirelessly from the supply network to electric cars.
- SAE J1772 [12]: In North America, this standard makes it easier for EVs and PHEVs to charge conductively.

In [6], four distinct EV charging modes are described. All charging modes require a residual current device.

- Mode 1: The most popular charging mode is Mode 1. It may operate in power systems that are single-phase or three-phase, with maximum VRMS values of 250 and 480, respectively. The highest discharging current permitted in this mode is 16 A, and no special connectors are required. The United States does not permit the use of this mode.
- Mode 2: This mode is utilized in power systems that are single-phase or three-phase, with maximum VRMS values of 250 and 480, respectively. An RMS current of at least 32 A is needed for the electric car in this mode, but there is no particular connector essential. This option requires an inline control box.
- Mode 3: With an electric vehicle (EV), the grid is connected through specialized supply equipment for electric vehicles (EVSE). The maximum RMS current that can be connected in single or three phases and be rated anywhere from 32 and 250 A.
- Mode 4: An off-board battery charger uses a DC connection to link the electric grid and the electric vehicle. In this mode 400 A is the maximum rated current.

Residential areas with household outlets charge the battery with full capacity in the slow charging Modes 1 and 2 [13]. Hotels, shopping centers and parking lots use Mode 3 as the semi-charging mode [14]. Mode 4 is the fast-charging mode that recharges the batteries within minutes which is used in highways, rest areas, and roads [15].

2.3 EV Charging Methods

Wireless charging, conductive charging, and battery exchange are the three primary charging techniques. Three further categories for conductive charging exist: overnight charging, top-down pantograph, and bottom-up pantograph. Figure 2.1 depicts various categories of electric vehicle charging techniques.

2.3.1 Battery exchange (BSS)

Battery swap station is the name of the battery exchange method where the battery is exchanged for a monthly rent payment to the proprietor of the battery swap station (BSS). The battery life is extended by using this slow-charging technique [16]. The batteries swap station and the generated renewable energy sources (RESs), such as solar and wind energy, can work together. The drivers do not have to replace the drained battery by getting out of their car to participate in the vehicle-to-grid (V2G) program with the reserve battery kept at the station.

The amount of rental charges to be paid to the BSS owner is high and is costlier than the fuelling process of the ICE engine. This method

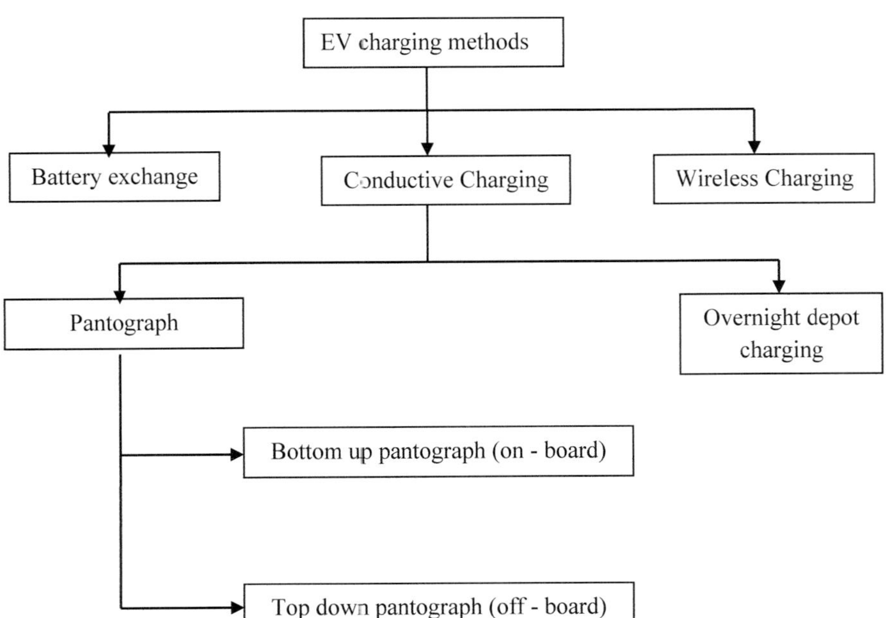

Figure 2.1 Classification of electric vehicle charging methods.

requires several high-priced batteries and a large area to store them, which is expensive. The station must accommodate different battery standards to suit the needs of vehicles [17].

2.3.2 Wireless power transfer (WPT)

The electromagnetic induction theory serves as the foundation for this innovative method. It makes use of two coils: the secondary coil is found within the car, while the primary coil is situated on the road. WPT technology's ability to recharge in a comfortable and safe manner makes it crucial for use in electric vehicle applications. The car can be charged while in motion, and a normal connection technology is not required [18]. Inductive power transfer is weak and requires an air separation between the transmitter and receiver coils of 20–100 cm for effective power transmission.

Eddy current loss occurs in wireless power transfer when the transmitter coil is not switched off. The transmitter and the EV would experience latency if they communicated in real time for data transfer [19]. The main benefit is that the recharging happens even when the car is moving and doesn't require a normal plug.

The conduction type of charging is based on supply types like DC and AC charging. AC charging, or slow charging, is the method of charging from a household plug point. The converter inside the electric vehicle converts the power into DC and delivers it to the car's battery. It takes about seven hours to fully charge a 40-kWh battery car. Electric vehicles fast charge using DC charging where power conversion takes place and the electricity is fed into the battery of the automobile. DC chargers have both low- and high-charging rate without limitation on their weight or size.

An electrical connection must be made between the car and the charging input in order to use conductive charging. Because there is a direct connection, it allows for various charging levels and has a high charging efficiency. A public charging station uses electricity charging levels two and three. The distribution system is less significantly impacted by the first and second layers (1 and 2). Conductive charging helps the V2G technology to minimize grid loss and avoid overloading the grid. Uncoordinated charging can lead to grid power overloading. The electrical distribution system may experience voltage instability as a result of rapid charging. With the assistance of the car's battery, CC may additionally afford reactive power correction while maintaining the voltage level [19].

The distribution system is impacted by the third level of power charging in a number of ways, including voltage variation [20], power loss, and system dependability. Peak demand rises and transformer life is shortened by the third power charging level [21]. The specifications include a standard connector with multiple charging levels, a sophisticated infrastructure, and controlled access to the power grid. The grid and vehicle must communicate rigorously in order to use V2G technology. Because batteries need to be charged and discharged frequently, it also shortens their lifespan.

2.3.3 Overnight depot charging

This type of charging is located at the end of the line and is utilized for charging during the night. Both quick and slow charging are acceptable. Slow charging is supported by the little impact of charging on the distribution grid [22, 23]. For quick battery charging, the pantograph charging method is suitable for larger battery capacities.

2.3.4 Pantograph charging

For devices like buses and vehicles that demand more power and have larger battery capacities, pantograph charging is employed. The bus battery must be inexpensive to use for this charging technique. When it comes to pantograph charging, bus investment costs are lower but infrastructure costs are higher. There are two categories for pantograph charging: bottom-up and top-down.

Top-down pantograph: Typically referred to as a top-down pantograph, this charging arrangement is installed on the bus stop's roof. Tests of this method's high power direct current output have been conducted in the US, Germany, and Singapore [24]. *Bottom-up pantograph*: In situations when the bus already has charging equipment installed, this method of charging is suitable. Another name for this is an on-board bottom-up pantograph [24].

2.3.5 Induction charging

The concept of electromagnetic induction and high frequency where energy is fed as input from the power source to the electric vehicle through magnetic induction coupling is called inductive charging. It uses two electromagnetically connected coils.

The power network is connected to the primary coils, which are located on the road surface and feature a pad-like design. Near the bottom of the car, at the bottom, is where the secondary coil is mounted. The AC is rectified and transformed into high frequency AC electricity at the charging station. After that, induction is used to transfer this power to the vehicle. Because battery temperature has a significant impact on battery life, a dependable charging protocol must be implemented to guarantee fast charging times and a safe battery pack temperature of less than 50 °C. An efficient charging standard must be maintained to achieve good capacity and high energy efficiency. Some of the important charging protocols are discussed below.

Constant current (CC) charge: In this method, the battery must receive a continuous charging current until its cell voltage reaches a threshold, at which point it slows down. Heat loss happens in the CC mode when the charge current is high [25].

Constant voltage charge: In this technique, the charging current gradually rises to a steady-state voltage that is equivalent to the battery voltage. When the battery voltage is reached, the current likewise gradually decreases. This method's key benefits are its controllability and quick charging time. The drawbacks include the inability to fully charge the battery and the fact that a high charging current would raise the battery's temperature, which will cause aging and damage [26].

Pulse charge: The battery can be charged quickly within a short duration using pulse current, but the quick-charging method can act as a disadvantage on the life of the battery [26].

Boost charging: Boost charging allows the battery charger to draw a large current for a brief period of time. This approach applies the 4C rate [27].

Ohmic drop compensation: This method takes into account the battery's Ohmic drop resistance with an initial increase in the preset cell voltage threshold. With this method, the maximum 6C rate can be attained.

Linearly declining current (LDC) charging: The predetermined state of charge (SOC) of the battery pack is used by the LDC to reduce the charging current linearly.

Multistage charging: Multi-stage charging involves three or more charging stages with multi-stage current, depending on the battery model.

2.4 EV Converter Topology

Nonlinear power load problem is a disadvantage of EV charger which creates low power factor and increase in the total harmonic distortion. An effective battery charger complies with international standards like IEEE 1547 and safely charges the battery pack. Power factor correction (PFC), isolation,

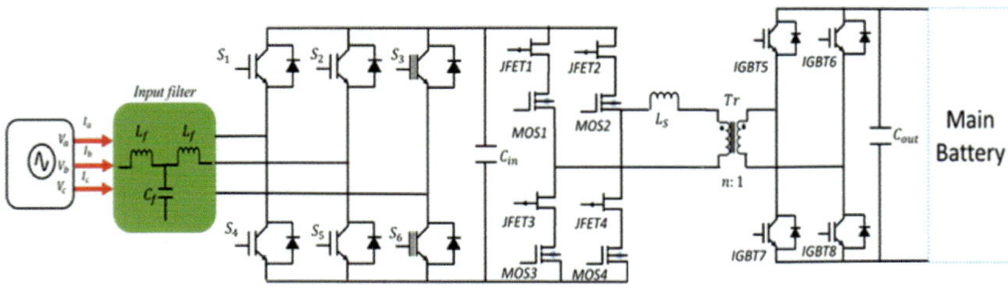

Figure 2.2 Three-phase on-board charger topology based on active rectifier and single DAB.

and filtering stages are all included in EV chargers. Various topologies of power convertors have been employed in order to increase battery efficiency. A single-phase PMW rectifier was employed in the first stage of the 6.6 kW prototype described in [28], while a full-bridge resonant converter was used in the second stage. Its rated efficiency is approximately 93%, and its power density is only 1 kW/L. This prototype employs a silicon (Si) super-junction Metal oxide semiconductor field effect transistor (MOSFET) in its power stage.

Based on a dual active bridge converter and a three-phase active rectifier, which have been shown to have a 91% efficiency under the least demanding battery charging conditions, a 10-kW three-phase prototype was shown in [29]. The three-phase on-board charger architecture based on a single dual active bridge (DAB) and an active rectifier is depicted in Figure 2.2.

2.4.1 Recent developments of DC–DC Converter

A modified DAB converter topology for an EV charger [30], uses a turned L-C-L network to reduce the reactive currents in the active bridges and with it the efficiency of the converter increased. This modified topology reduces the number of capacitors compared to the conventional DAB topology, but the inclusion of the L-C-L filter increases the converter volume. In the CLLC-based series-resonant converter (SRC) topology for EV charging applications, the DAB converter design is less complicated than CLLC resonant network design (Figure 2.3).

Numerous factors, including the secondary transformer voltage, the load situation, the inductance values, and the capacitance of the capacitors employed in the CLLC network, might affect the voltage gain in this design [31, 32]. Compared to the DAB architecture, the voltage gain equation in the CLLC SRC is more complicated. Several CLLC parameters must be tuned for

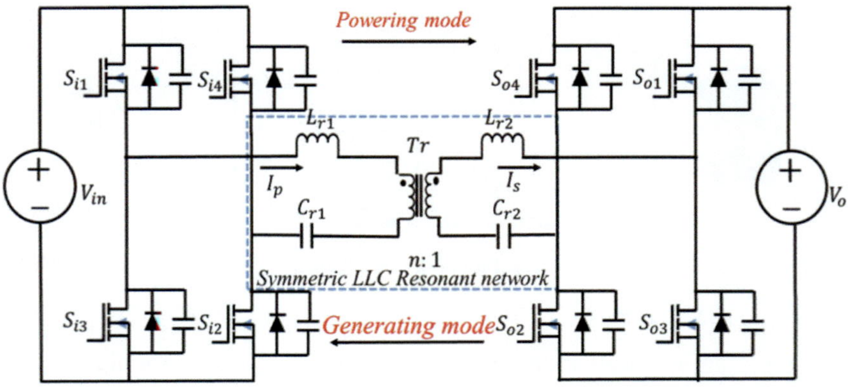

Figure 2.3 SRC DC–DC converter topology.

the EV charger's charging and discharging modes in order to enhance system performance. When compared to the DAB topology, this makes the design process more difficult.

To lower the reactive current in the active bridges, a variety of sophisticated modulation techniques are employed, including phase-shifting pulse width modulations (PWM), trapezoidal modulations, and triangle modulations. Additionally, the DAB topology's power efficiency is increased by these modulation approaches. In similar bridge switches, reactive current reduction is achieved by combining dual PWM with pulse shift modulation [32]. To increase the topology's power efficiency, separate switching devices are independently controlled by pulse width modulation (PWM) using a composite modulation technique [33].

In composite modulation, the root mean square (RMS) current is reduced in each switching device. On the other hand, it requires complex control algorithm, and the modulation strategy needs modification at various load conditions.

For EV fast chargers, the dual active half-bridge (DAHB) converter is a crucial DC–DC topology (Figure 2.4). Compared to the DAB topology, this architecture features fewer semiconductor switches. It is challenging to control closed loop stability in DAHB due to the existence of low-frequency resonance. Additionally, typical phase shift modulation has a very narrow ZVS range. To overcome the stability problems and increase the range of zero voltage switching (ZVS), an enhanced asymmetrical modulation technique is required.

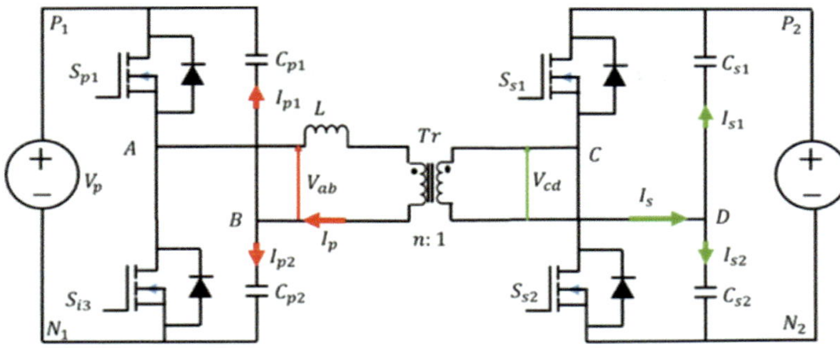

Figure 2.4 DAHB convertor topology.

2.4.2 Recent developments of AC–DC converter

The two-level active PWM rectifiers and the three-phase diode rectifier with unidirectional topology are the most often utilized topologies for the AC–DC converter stage. The three-level phase-leg-based neutral point clamped (NPC) AC–DC converter topology minimizes the requirement for filtering during the EV charger's power factor adjustment phase (Figure 2.5).

The pulse width modulation strategy and the three-level operation involved in this topology can reduce the input filter size. In this topology, the number of semi-conductor devices required is increased two times when compared with a two-level active rectifier, whereas the voltage rating of each semiconductor device is reduced, enabling the power density of chargers greater than 10 kW [34].

The 11 kW chargers achieve 2.5 kW/L power density and 95.6% peak efficiency by a three-level NPC PWM boost rectifier charger architecture that incorporates a revolutionary phase shift modulation scheme and multi-port design. The primary drawback of this topology is the three-fold increase in semi-conductor devices when compared to two-level rectifiers [35].

Applications for fast charging do away with the requirement for a traditional AC–DC arrangement. In Figure 2.6, the topology diagram is displayed. The three-phase 50/60 Hz AC voltage is converted into a single-phase high-frequency AC voltage by the front-end matrix converter. A high-frequency transformer creates the galvanic isolation, while a pulse width modulation rectifier converts the high-frequency AC voltage into DC. The characteristics of this topology include a single-stage power conversion and the removal of the DC link capacitor at the output of the typical AC–DC stage. As a result,

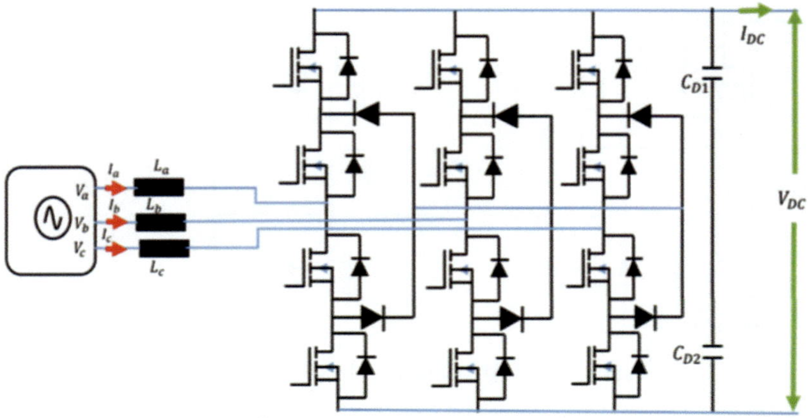

Figure 2.5 Three-level NPC boost AC–DC converter topology.

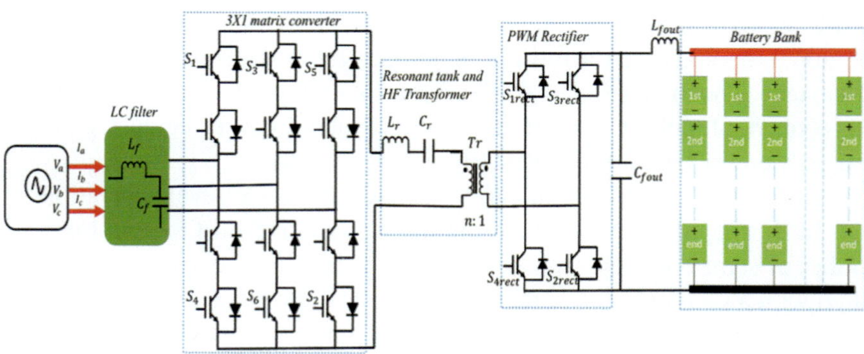

Figure 2.6 Matrix converter topology with resonant tank and HF transformer for EV charger.

compared to other AC–AC converter types, matrix converters attain a higher power density [36].

2.5 EVs Battery Chargers Architectures

There are two cascading stages on the on-board single-phase battery chargers (Figure 2.7). An AC-DC converter that pulls low harmonic sinusoidal current from the grid is the first stage. This stage complies with IEC 61000-3-2 standard and contains a unity power factor correction (PFC). A DC–DC converter in the second stage controls the battery's current according to its state of charge (SOC). It balances the voltage differential between the battery and the DC-link.

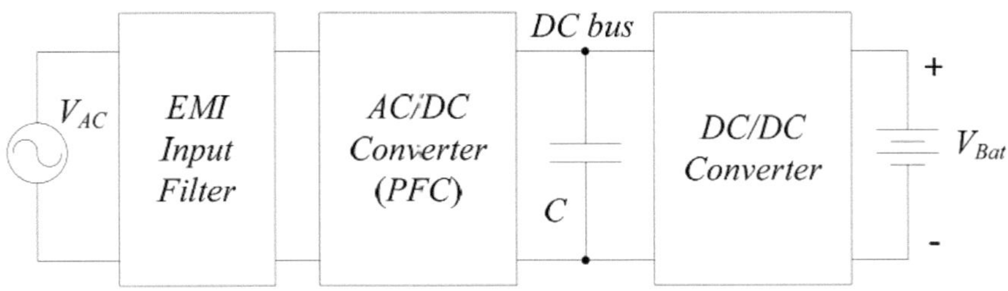

Figure 2.7 General block diagram of a single-phase on-board battery charger.

The two stages are interconnected by a DC-link capacitor. As per the International Special Committee on Radio Interference (CISPR) 22 standard, the interconnection between the grid and first stage has an electromagnetic inference (EMI) filter.

2.5.1 Power factor correction (PFC) stage circuit topologies

A boost converter comes after a diode rectifier in a unidirectional AC–DC converter. Its input current is continuous and its structure is simple. The primary drawback is the conduction loss that the diode rectifier produces.

To overcome this, various bridgeless topologies with high efficiency are used that restrict the use of a rectifier, as shown in Figure 2.8(a)–(c) [37].

Parallel connection of various converters is an appropriate method to overcome the increase in power level without significant increase in the current stress on the components, as shown in Figure 2.9 (a).

The filtering components' size and the ripple in the current are reduced by the converters that are connected in parallel and operate through interleaving. This method is also suitable for bridgeless topologies, as observed in Figure 2.9(b) [38].

To enable bidirectional power flow in a unidirectional boost converter, the diode rectifier can be swapped out for a synchronous rectifier. To achieve increased efficiency, a bidirectional boost converter can be used in place of the synchronous rectifier's single leg as shown in Figure 2.10.

2.5.2 Battery current regulation stage circuit topologies

In order to lower the voltage from the DC link to the battery voltage level, a unidirectional buck convertor is part of the battery current regulation (BCR) step of a DC–DC converter. Batteries for electric vehicles typically

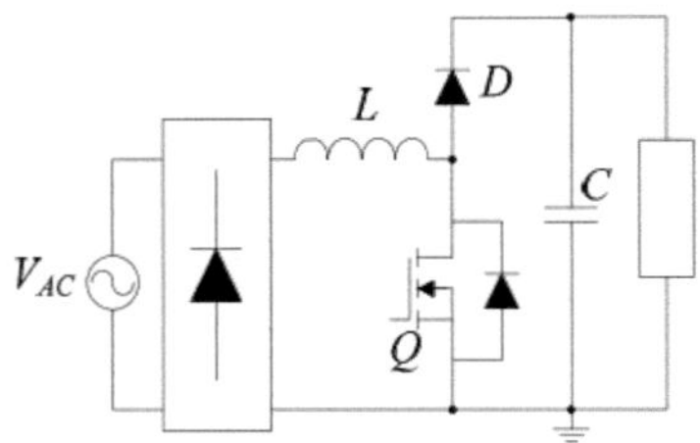

(a) Diode rectifier and boost converter.

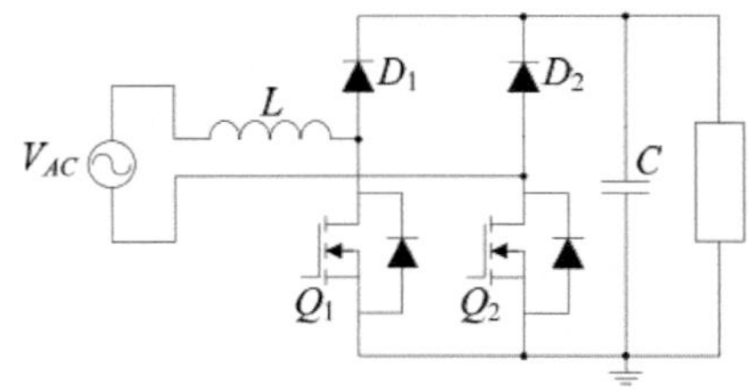

(b) Basic bridgeless topology.

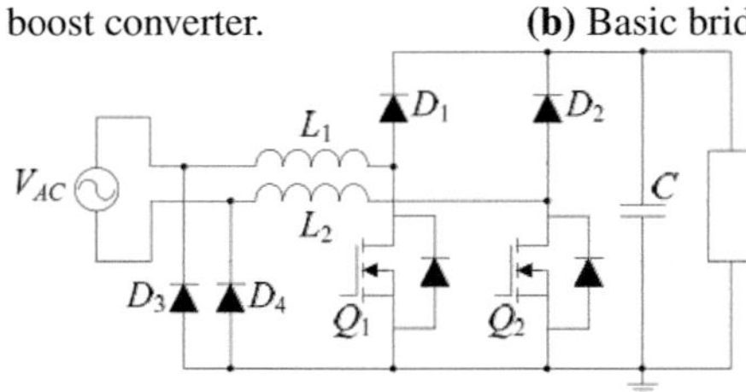

(c) Dual-boost or semi-bridgeless topology [41].

Figure 2.8 Unidirectional boost-based AC–DC converters.

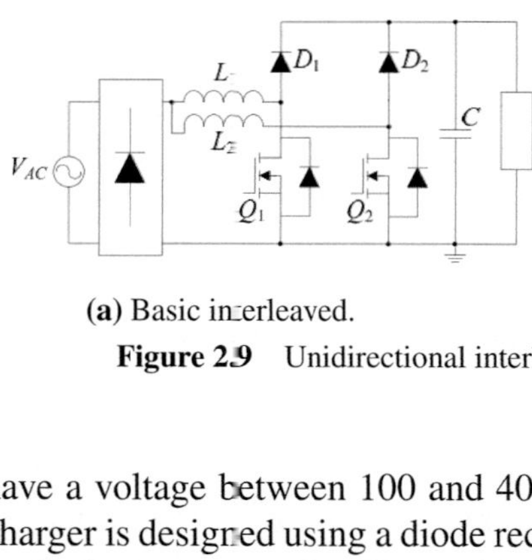

(a) Basic interleaved.

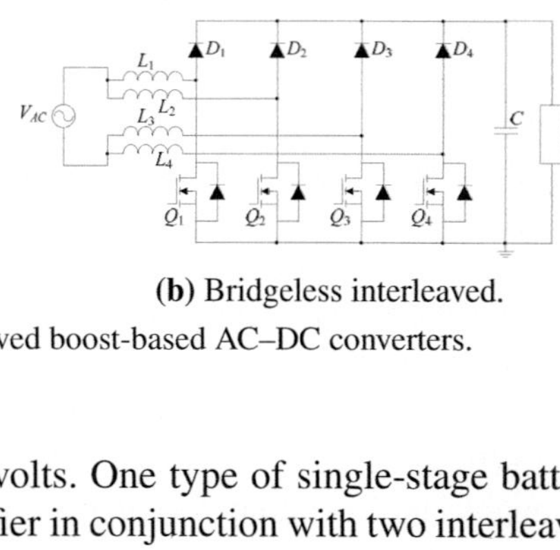

(b) Bridgeless interleaved.

Figure 2.9 Unidirectional interleaved boost-based AC–DC converters.

have a voltage between 100 and 400 volts. One type of single-stage battery charger is designed using a diode rectifier in conjunction with two interleaved buck–boost converters (Figure 2.11) [39].

A high-frequency switching transformer is employed in some circuit topologies of the battery current control stage to enable galvanic separation between the grid and the battery. Compared to line frequency transformers, high frequency switched transformers have advantages in terms of weight and size. The two most popular bidirectional topologies are the full-bridge series resonant converter and the phase-shifted full-bridge DC–DC converter, as illustrated in Figure 2.12 (a).

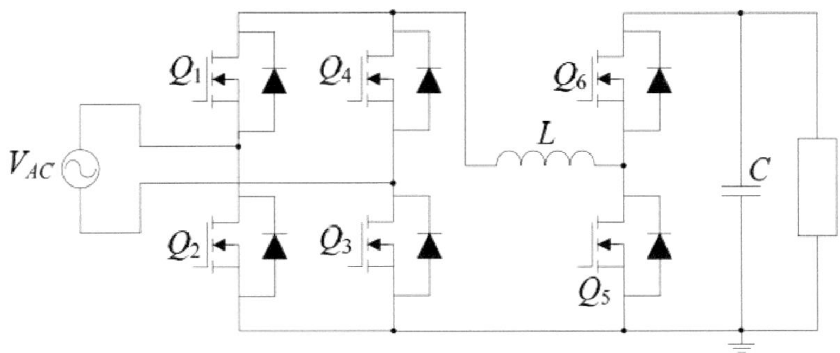

Figure 2.10 Bidirectional boost-based converters with synchronous rectifier.

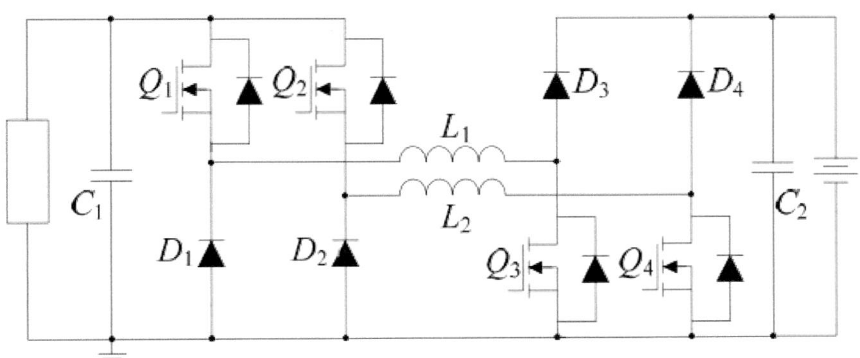

Figure 2.11 Unidirectional nonisolated interleaved buck–boost DC–DC converters.

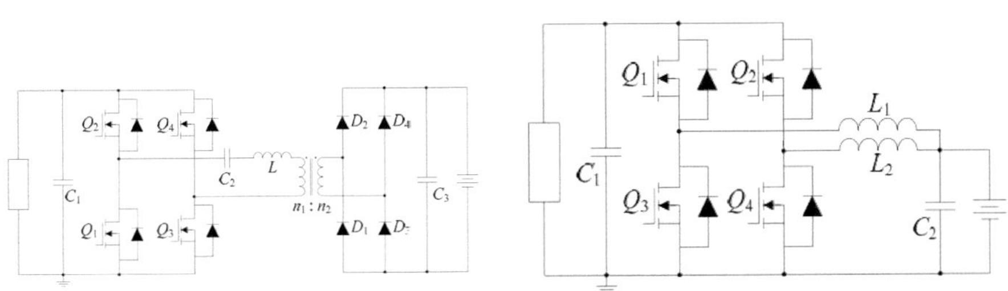

(a) (b)

Figure 2.12 **(a)** Unidirectional isolated full-bridge series interleaved two quadrant buck DC–DC converters. **(b)** Bidirectional nonisolated resonant DC–DC converters.

A bidirectional buck converter, as shown in Figure 2.12 (b), is used to supply energy to the DC-link in a traction power system to control the electrical drive of the vehicle [40].

Dual active bridge and bidirectional full-bridge series resonant converters are used in EV battery chargers with bidirectional power flow capacity and galvanic isolation.

2.6 Battery Charger Controllers

The two-stage battery charger design uses one controller for each stage, as shown in Figure 2.13.

The input voltage and current are proportionate to the PFC controller. It attains a power factor of one. At a specific reference voltage, the PFC controller controls the DC link voltage. An outer loop regulates the DC-link voltage, whereas an inner loop controls the input current in single-phase PFC controllers. To find the inner loop's reference, multiply the outer loop's output by the input voltage that was measured.

To prevent the grid voltage being sensed under contaminated conditions the phase lock loop system is used [40]. To reduce the low-frequency harmonics added into the grid current the bandwidth is chosen in the range of 1020 Hz for the outer loop. The output of the DC link voltage compensator is subject to oscillations at twice the line frequency. To prevent this phenomenon notch filters are used in the outer loop [41].

The battery current in the battery charger controller's inner loop is controlled based on the reference current produced by the battery management system (BMS). The battery's state of charge is used to determine the battery

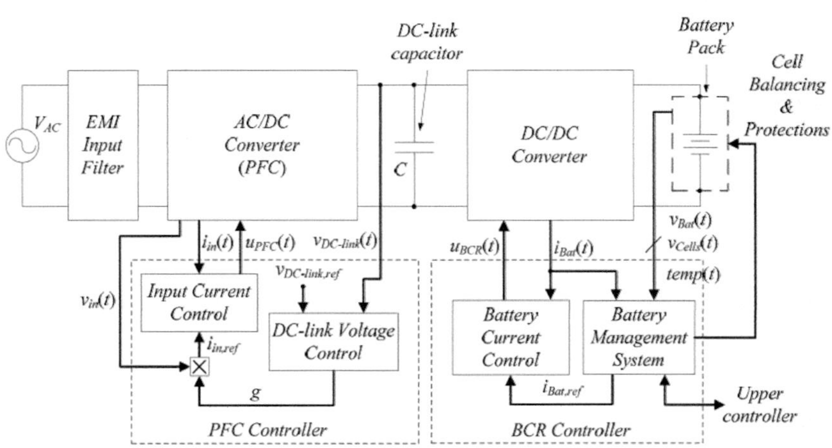

Figure 2.13 Block diagram of a battery charger controller.

reference current. The temperature and voltage of the battery, among other factors, are used to estimate the state of charge (SOC). The BMS is dependable for balancing the cells that make up the battery pack and is connected to a higher controller.

The constant current, constant voltage (CC-CV) method is the most commonly used method to charge a battery.

This approach involves charging the battery continuously at first (CC) until the voltage reaches a predetermined level. The battery charger enters constant voltage (CV) mode when the preset setting is reached, which causes the SOC and battery current to progressively drop.

The microprocessor-based control unit has led to the growth of other techniques like the genetic algorithm that finds the optimal value of charging current in accordance with the battery model. The sinusoidal current is amplitude and frequency modulated to find the lowest input impedance to enable efficient charging operation in a fuzzy controller. The high cost and complexity involved in analog controllers have favored digital controllers for system management. The digital controllers have the advantages of low cost, small size, and enlarged computation capability.

An analog and digital controller combined with a two-stage battery charger is employed to enable high sampling rates and different control algorithm implementation for power factor adjustment.

There are two types of inductor current mode controllers: those with a constant switching frequency and those with a variable switching frequency. A self-oscillating current mode controller with a configurable switching frequency is the basis of the hysteretic current mode control technique [42].

The internal change of state in the converter determines the type of topology. The dynamic response of the hysteresis-based switching converter is depicted by efficiently most suitable method called sliding mode control (SMC) approach. SMCs enable the synthesis of loss-free resistors (LFRs).

The input voltage and current of a sliding mode converter are proportionately related. An LFR is made out of a two-port structure with a voltage-proportional input current. The fact that there is no power loss during transmission makes LFRs a kind of POPI (power output = power input) system.

LFRs are used in power factor correction and impedance matching applications. The advantages of hysteretic current-mode controllers are fast dynamic response, stability, and robustness. The disadvantages are the complex design of the filtering elements due to variable switching frequency, and the interleaving method is restricted to specific configurations. In constant frequency-based current controllers, a sliding mode approach with constant switching frequency provides a fast dynamic response and is suitable for PFC applications [43]. The constant switching frequency is a direct application of the interleaving method.

The average, peak, and valley current-mode controllers for boost, buck, and buck–boost power converters are among the predictive current controllers with constant switching frequency. Using the input and output voltages, the inductor current, the plant parameters, and the present duty cycle, the predictive controllers calculate the duty cycle of the upcoming switching period.

It has been established in [44] that boost converters like discrete time proportional-integral (PI) current mode controllers, which are a class of predictive current mode controllers, exhibit faster dynamic response and better input current characteristics.

One benefit of using digital controllers is their adaptability and capacity to incorporate intricate algorithms. They are extensively utilized in battery SoC monitoring, battery charging applications, and system communication. With a quicker dynamic response, hybrid controllers are perfect for controlling capacitor voltages and inductor currents.

2.7 Electric Vehicle Charging Challenges and Possible Solutions

The grid receives power from bidirectional chargers. The system's safety may be compromised by common mode noise between the electric car and the grid as well as by improper grounding. It is possible to add a transformer to the converter in order to eliminate common mode noise. The weight of the entire converter increases with the inclusion of a transformer.

A transformerless topology is useful to reduce the common mode noise, but this increases the system cost due to the inclusion of a device to avoid the freewheeling current. The presence of the electric vehicle can have negative effects on the grid distribution like voltage fluctuations, harmonic distortion, and overheating of the distribution transformers.

A rapid charging station that can charge in less than 30 minutes could be a viable way to get around the slow charging rate, which takes 7–8 hours to complete. However, this leads to problems with grid distribution, and in order to maintain grid stability, an energy buffer unit is needed.

New power converter topologies are essential to deliver fast charging in electric vehicles. The fast-charging station must ensure soft switching based on low power loss, safe charging capacity in less time, provision for bidirectional power flow, a separate unit for driver safety, and resonance attainment for maximum efficiency. The converter design must focus on high power pulse charging so that the battery management system can ensure battery strength and control the charging process.

Filter design and power factor correction topologies are required to control the excessive charging current that strains the DC bus and interferes with the smooth operation of the electric grid. A fast-charging system must be adaptable enough to work with multiple battery types from different producers. Reducing converter loss and improving thermal stability in a fast-charging system is possible with the use of broadband devices. For electric vehicles, the most popular power source is the lithium-ion battery. Battery technology has a number of difficulties, including heat buildup inside the battery pack, constant load, and regenerative braking.

Space limitations lead to thermal difficulties that deteriorate battery performance. The heat source model determines how accurate the battery temperature is estimated to be. The reversible model experiences heat loss due to the chemical reaction's enthalpy shift, while the joule heating heat source model exhibits an ohmic drop in the battery's internal resistance [45–46]. The battery pack may be harmed by battery thermal runaway, which is brought on by abnormal heating. Several thermal management strategies (TMSs) have been put forth in recent studies to ensure the battery operates efficiently.

There are two types of TMS cylindrical Li-ion batteries: liquid and air. For an air TMS, a wide battery module with a small spacing between cells is appropriate, but a liquid TMS benefits from a narrow battery module with a tiny gap. The air TMS consumes more power than the liquid TMS when there is a significant heat load. A Li-ion battery is utilized in forced air conditioning systems.

Various heat transfer methods like convection and advection are used to access the cooling performance in heat management. The reduction in hydraulic diameter and increase in coolant flow rate improve cooling performance. These factors in turn increase the operating power of the fan. In forced liquid cooling, type TMS bipolar lithium-ion batteries are used.

The coolant plate thickness and high coolant velocity exert a strong control over the maximum temperature. The temperature is kept constant by these elements. The average temperature rises due to the quantity of cells between the cooling plates and increasing discharge rates. The coolant velocity is increased in order to lower the average temperature.

The nonuniform temperature is maintained with high coolant plate thickness and coolant velocity. There exist an inverse relationship between the average temperature and coolant velocity. In phase change materials (PCMs), type TMS Li-ion batteries are used.

PCM lowers the cell's maximum temperature as well as its temperature spread. Higher PCM thickness and a higher degree of curvature around the cell provide improved cooling within the cell. The TMS prismatic Li-ion battery is utilized in air conditioning. The fan's gap spacing is lowered and the flow rate is increased to limit the maximum temperature growth. Although the unequal spacing influences heat circulation, it has no effect on the maximum temperature growth.

Constant gap spacing reduces both the maximum temperature growth and the overall temperature uniformity. Passive TMS can achieve a consistent temperature under typical circumstances. Thermal runaway is avoided by heat absorption through the PCM graphite matrix. Li-ion batteries made by TMS are utilized in heat pipe cooling. The thermal resistance of the heat sink is decreased by the addition of the heat pipe. Under a variety of circumstances, flat heat pipe functions efficiently. The heat pipe efficiently increases the heat flux more than the traditional heat sink when operating at high frequencies. Li-ion batteries made by TMS are utilized in wet and heat pipe cooling systems. This kind of thermal management system does not recommend the use of a natural convection cooling system.

The battery discharges faster due to the high battery temperature and temperature gradient value. It is not recommended to cool heat pipe TMS using water bath since bubbles will form during discharge. A Li-ion battery powers a TMS of the liquid cooling variety. The increasing inlet mass circulation sets a maximum temperature limit. The breadth of the cooling plate, the inlet mass flow rate, and the inlet temperature are inversely proportional to each other [47–21].

2.8 Smart Battery Charger

Another name for smart battery chargers is microprocessor-controlled chargers. Smart EV batteries are compatible with these chargers. Special microchips included with smart EV batteries are intended to communicate with charger produced by the same manufacturer. Dumb chargers are less sophisticated than smart chargers. It must be plugged in and charged. Charging sessions are managed by the vehicle app. Dumb chargers has a maximum single-phase charging power of 7.4 kW. They will maintain charging the automobile all day long and have no authority to stop. In addition to having timed charging intervals and solar integration for charging a car using solar panels, smart chargers offer complete control over the charging process.

Dumb chargers have the disadvantage of not collecting data on energy usage. It is feasible to plan charging and discharging for a specific day and day ahead of time with smart charging. The Vehicle-to-grid (V2G) is compatible with smart chargers. The other charging topologies are not suitable for V2G because they are regarded as unmanageable loads. Since battery electric vehicles do not produce any direct or indirect pollutants, the Environmental Protection Agency classifies them as zero-emission vehicles. A battery-electric vehicle's maximum speed is determined by the battery's capacity, and charging does not cease until the batteries are fully charged.

Electrical energy systems known as aggregators serve as a conduit between power systems and end users. Aggregators have the ability to affect several grid-connected units. Enhancing power system control and energy program optimization is the primary goal of aggregators. Demand response aggregators, retail aggregators, and electric vehicle aggregators are the many categories of aggregators. Customers who participate in demand response programs receive contracts for consumption reduction from the demand response aggregators [52]. As EV integrators become increasingly prevalent in power systems, researchers are focusing more on understanding their behavior. By integrating EVs as distributed energy sources, EV aggregators serve as an interface and assist the V2G protocol. When managing the battery state-of-charge assessment of geographically dispersed EVs, EV aggregators can be useful [53].

EV aggregators are also accountable for electricity supply and electric charging and discharging control in a particular area. The framework of EV with aggregators is shown in Figure 2.14.

The transmission system operator, distribution system operator, and energy market suppliers are coordinated by EV aggregators, which also produce vital signals. Customers with roof-top solar power systems can use vehicle-to-home (V2H) integration to supply electricity during peak hours or blackouts. When EVs are used in conjunction with renewable energy sources, they are able to absorb more energy and utilize it for transportation. There is a huge increase in the amount of renewable energy sources to lessen consumer dependence on fossil fuels. A micro grid is created by producing energy sources in the form of micro generators.

Micro grids provide intelligent charging of electric vehicles (EVs) to optimize energy efficiency and boost dependability. Therefore, in micro

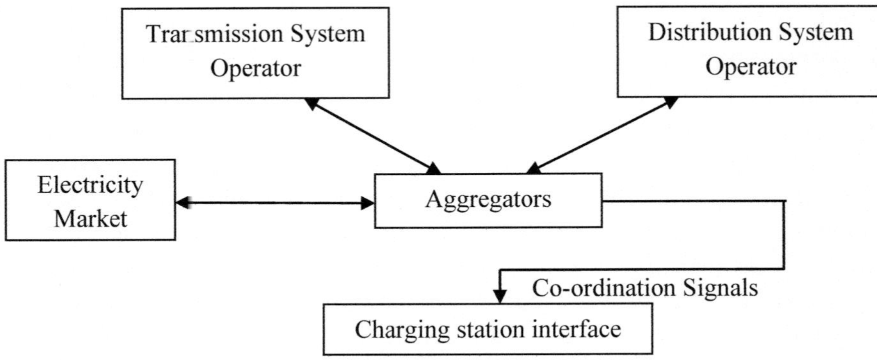

Figure 2.14 Frame work of EV with aggregators.

grid systems, combining renewable energy sources with EVs will have a significant role in controlling energy usage, cutting costs, and lessening environmental effect. Distributed energy systems may be used in the future to charge electric cars that are connected to residences. Since electric vehicles charge and discharge at the appropriate times, they reduce peak demand and smart home electricity bills. Thus, EV performance maximizes the total efficiency of smart networks and smart houses. Research on controlling energy use emphasize the integration of electric vehicles (EVs) with homes that have energy storage, heat, power, and wind turbines.

A lot of research is being done to combine PHEVs and smart buildings. Optimizing energy consumption control and boosting power supply reliability are the goals of combining these two exciting technologies [54]. Multi-carrier energy systems with the capacity to transfer, convert, and store various carriers are called energy hubs. Energy consumption management is necessary when EVs are incorporated into energy hubs since this throws off the demand–supply balance. A network of dispersed, medium-sized power-generating units and adaptable storage devices is known as a virtual power plant. EVs enhance the virtual power plant's frequency response and economic performance. Wind turbines and other renewable energy sources can be combined to operate as a virtual power plant for EVs.

2.9 Vehicle to Grid (V2G)

An effective energy-consumption plan is required for electric vehicles. Since EVs operate as a load when charging, an overload may result from a rise in the total number of vehicles. Electric vehicles (EVs) can be used as loads and energy storage systems (ESSs) in vehicle-to-grid (V2G) technology, which supports the power grid. As a result, V2G is a useful technology that offers an answer for the big number of electric cars (Figure 2.15).

The functionality of EVs has been simplified with the introduction of vehicle-to-grid (V2G) technology, which permits bidirectional energy transmission between EVs and the electrical grid. The power supply in a V2G network is significantly impacted by uncoordinated EV charging and discharging. There are several different EV charging schedule schemes that can be used to lower the costs related to the charging process and enhance grid reliability.

The distributed operation features and great scalability are provided by the hybrid charging and discharging process. As a result, plug-in electric vehicles (PEVs) require less of a dispersed charging system and a centralized regulated scheme [55]. In a smart grid, consumers can customize the PEV charging rates to suit their needs, increasing the overall structural efficiency [56].

Using a range of charging and discharging processes can help reduce the impact on the utility system, boost cost-effectiveness, predict correct charge bills, and limit the amount of electricity used. The unpredictable nature of the electricity charge can be represented as a Markov decision process (MDP) with a transition possibility from the customer's point of view. To optimize the owner's income by providing the best power charge, it is essential to estimate the charging and discharging behaviors of EV batteries.

V2G technology facilitates bidirectional energy transmission between the electric car and the power grid. If the EV's battery has a large energy storage capacity, its energy is transferred to the power grid. If the electric car's battery

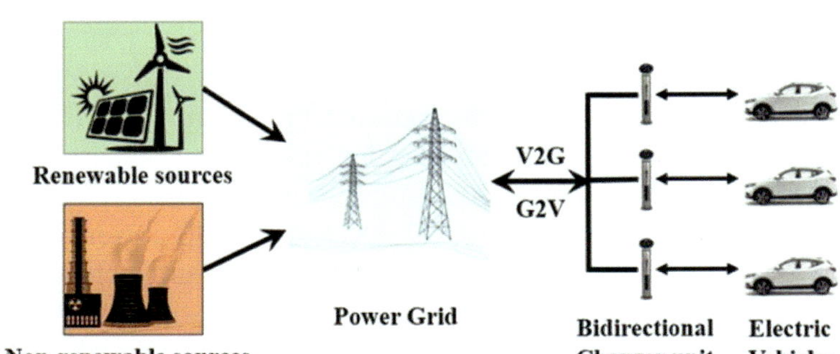

Figure 2.15 Vehicle-to-grid framework.

life is short, electrical energy flows from the power grid to it. EVs that use the V2G approach are usually charged during periods of higher energy production or lower power costs.

When there is a need for electricity or during peak load hours, they restore the energy to the power system. Reactive power compensation, compensation for renewable energy sources, and subsidiary services are only a few of the benefits of V2G. Because EVs may support subsidiary services, the power grid operates more steadily and is less susceptible to unplanned events.

EVs can supply power or act as a load during the frequency regulation procedure. Compensation from renewable energy sources facilitates spinning reverse services, load leveling, and frequency adjustment [57]. The energy stored within EV batteries can be sold for a profit with the help of V2G technology, which benefits EV owners. It also lessens the payback period, which counteracts the electric vehicle's high initial cost. The system operator is guaranteed prospective grid facilities via the V2G strategy. Reactive power adjustment is provided by the V2G technology's bidirectional battery charger without the need for any active power exchange [58].

In order to reduce distribution system losses, V2G technology uses two scheduling strategies: reactive power dispatch (RPD) and active power dispatch (APD). The RPD approach uses the best reactive power addition to minimize losses, whereas the active power dispatch (APD) uses the V2G scheme to minimize inefficiencies in the distribution system.

2.10 Conclusion

One potential way to lessen the negative environmental effects of carbon emissions from internal combustion engine-powered automobiles is to electrify them. Numerous studies have looked into ways to solve problems with rapid charging that have an impact on charging times. New charging technologies and power converters that provide an EV battery with a dependable charging method are required as the number of electric vehicles on the road rises. Adoption of coordinated charging systems and smart charging infrastructure for electric vehicles (EVs) can lead to optimal efficiency in vehicle-to-grid integration. Studies on DC–DC and AC–DC power converters for EV charging systems and various charging techniques are included in this chapter. Furthermore, the architecture of battery chargers for electric vehicles is also explained, along with the many charging issues and potential solutions. Some of the more modern technologies that improve energy consumption control and grid stability, such as smart battery charging and car to grid technology were also elaborated. Future research must examine the implications of fast charging on battery health and aging, even if a lot of work has been done to address fast-charging issues. Further research is needed to understand the process of battery failure caused by high current density and the right thermal challenge. EV wireless charging and energy networks are the foundation for the advancement of EVs and energy structures in the future. Future study must focus on EV charging in order to lessen the negative effects of EV integration on the grid. To overcome the limitations of power grid and battery function and improve the technological performance of power grid function, an ideal power flow based on EV charging and discharging strategy is needed. Grid stability is maintained using a number of control strategies to improve grid ancillary services, such as reactive power production for grid voltage support, harmonic correction, and fault current injection in the event of grid-side problems. The addition of EV charging stations to grid ancillary services would provide additional revenue for the station operators in the event that new restrictions are implemented.

References

[1] Lee, J.Y.; Chae, H.J. 6.6-kW On board Charger Design Using DCM PFC Converter With Harmonic Modulation Technique and Two-Stage DC/DC Converter. IEEE Trans. Ind. Electron. **2013**, 61, 1243–1252.
[2] Amjadi, Z.; Williamson, S.S. Power-Electronics-Based Solutions for Plug-in Hybrid Electric Vehicle Energy Storage and Management Systems. IEEE Trans. Ind. Electron. 2009, 57, 608–616.
[3] Rafi, A.H.; Bauman, J. A Comprehensive Review of DC Fast-Charging Stations with Energy Storage: Architectures, Power Converters, and Analysis. IEEE Trans. Transp. Electrification 2020, 7, 345–368.
[4] C. C. Chan, “The state of the art of electric, hybrid, and fuel cell vehicles,” Proceedings of the IEEE, vol. 95, no. 4, pp. 704-718, Apr. 2007.
[5] Zeraoulia, M., M.E.H. Benbouzid and D. Diallo, 2006. Electric motor drive selection issues for HEV propulsion systems: A comparative study. IEEE T.Veh. Technol., 55(6): 1756-1764.
[6] Arif, S.M.; Lie, T.T.; Seet, B.C. A Novel Simulation Model for analyzing the State of Charge of Electric Vehicle. In Proceedings of the IEEE PES, Innovative Smart Grid Technologies Asia, Singapore, 22–25 May 2018; pp. 1–5.

[7] Das, H.S.; Tan, C.W.; Yatim, A. Fuel cell hybrid electric vehicles: A review on power conditioning units and topologies. Renew Sustain. Energy Rev. 2017, 76, 268–291.
[8] Zhou, Y.; Huang, L.; Sun, X.; Li, L.; Lian, J. A Long-term Energy Management Strategy for Fuel Cell Electric Vehicles Using Reinforcement Learning. Fuel Cells 2020, 20, 753–761.
[9] Electric vehicle conductive charging system - Part 1: General requirements, IEC 618511, 2010.
[10] Plugs, socket-outlets, vehicle connectors and vehicle inlets - Conductive charging of electric vehicles – Part 1: General requirements, IEC 62196-1, 2014.
[11] Electric vehicle wireless power transfer (WPT) systems - Part 1: General requirements, IEC 61980-1, 2015.
[12] Electric Vehicle and Plug in Hybrid Electric Vehicle Conductive Charge Coupler, SAE J1772, 2012.
[13] D. Aggeler, F. Canales, H. Zelaya-De La Parra, A. Coccia, N. Butcher, and O. Apeldoorn, "Ultra-fast DC-charge infrastructures for EV-mobility and future smart grids," IEEE Smart Grid Technologies Conference Europe (ISGT Europe), Gothenburg, Oct. 11-13, 2010, pp. 1-8.
[14] M. Yilmaz and P. T. Krein, "Review of battery charger topologies, charging power levels, and infrastructure for plug-in electric and hybrid vehicles," IEEE Trans. Power Electronics, vol. 28, no. 5, pp. 2151-2169, May 2013.
[15] Arif, S.M.; Lie, T.T.; Seet, B.C.; Ayyadi, S.; Jensen, K. Review of Electric Vehicle Technologies, Charging Methods, Standards and Optimization Techniques. Electronics 2021, 10, 1910.
[16] Ahmad, A.; Khan, Z.A.; Alam, M.S.; Khateeb, S. A Review of the Electric Vehicle Charging Techniques, Standards, Progression and Evolution of EV Technologies in Germany. Smart Sci. 2017, 6, 36–53.
[17] Erdinç, O.; Taÿscikarao glu, A.; Paterakis, N.G.; Dursun, I.; Sinim, M.C.; Catalão, J.P. Comprehensive optimization model for sizing and siting of DG units, EV charging stations, and energy storage systems. IEEE Trans. Smart Grid 2017, 9, 3871–3882.
[18] Chowdhury, S.R. A Three-Phase Overlapping Winding Based Wireless Charging System for Transportation Applications; University of Akron: Akron, OH, USA, 2021.

[19] Patil, D.; McDonough, M.K.; Miller, J.M.; Fahimi, B.; Balsara, P.T. Wireless Power Transfer for Vehicular Applications: Overview and Challenges. IEEE Trans. Transp. Electrif. 2018, 4, 3–37.
[20] Dharmakeerthi, C.H.; Mithulananthan, N.; Saha, T.K. Impact of electric vehicle fast charging on power system voltage stability. Int. J. Electr. Power Energy Syst. 2014, 57, 241–249.
[21] Habib, S.; Khan, M.M.; Abbas, F.; Sang, L.; Shahid, M.U.; Tang, H. A Comprehensive Study of Implemented International Standards, Technical Challenges, Impacts and Prospects for Electric Vehicles. IEEE Access 2018, 6, 3866–13890.
[22] Arif, S.M.; Lie, T.T.; Seet, B.C.; Ayyadi, S. A novel and cost-efficient energy management system for plug-in electric bus charging depot owners. Electr. Power Syst. Res. 2021, 199, 107413.
[23] Meishner, F.; Satvat, B.; Sauer, D.U. Battery electric buses in european cities: Economic comparison of different technological concepts based on actual demonstrations. In Proceedings of the 2017 IEEE Vehicle Power and Propulsion Conference (VPPC), Belfort, France, 11–14 December 2017; pp. 1–6.
[24] Carrilero, I.; González, M.; Anseán, D.; Viera, J.C.; Chacón, J.; Pereirinha, P.G. Redesigning European Public Transport: Impact of New Battery Technologies in the Design of Electric Bus Fleets. Transp. Res. Procedia 2018, 33, 195–202.
[25] Vu, V.-B.; Tran, D.-H.; Choi, W. Implementation of the Constant Current and Constant Voltage Charge of Inductive Power Transfer Systems With the Double-Sided LCC Compensation Topology for Electric Vehicle Battery Charge Applications. IEEE Trans. Power Electron. 2017, 33, 7398–7410.
[26] Bayati, M.; Abedi, M.; Farahmandrad, M.; Gharehpetian, G.B. Delivering Smooth Power to Pulse-Current Battery Chargers: Electric Vehicles as a Case in Point. IEEE Trans. Power Electron. 2020, 36, 1295–1302.
[27] Notten, P.; Veld, J.O.H.; van Beek, J. Boostcharging Li-ion batteries: A challenging new charging concept. J. Power Sources 2005, 145, 89–94.
[28] Lee, B.-K.; Kim, J.-P.; Kim, S.-G.; Lee, J.-Y. An Isolated/Bidirectional PWM Resonant Converter for V2G (H) EV On-Board Charger. IEEE Trans. Veh. Technol. 2017, 66, 7741–7750.
[29] Zeljkovic, S.; Vuletic, R.; Miller, A.; Denais, A. A three phase bidirectional V2G interface converter based on SiC JFETs. In Proceedings of

the 2015 17th European Conference on Power Electronics and Applications (EPE'15 ECCE-Europe), Geneva, Switzerland, 8–10 September 2015; pp. 1–10.

[30] Twiname, R.P.; Thrimawithana, D.J.; Madawala, U.K.; Baguley, C.A. A New Resonant Bidirectional DC–DC Converter Topology. IEEE Trans. Power Electron. 2013, 29, 4733–4740

[31] Deng, J.; Li, S.; Hu, S.; Mi, C.C.; Ma, R. Design Methodology of LLC Resonant Converters for Electric Vehicle Battery Chargers. IEEE Trans. Veh. Technol. 2013, 63, 1581–1592.

[32] He, P.; Khaligh, A. Comprehensive Analyses and Comparison of 1 kW Isolated DC–DC Converters for Bidirectional EV Charging Systems. IEEE Trans. Transp. Electrif. 2016, 3, 147–156.

[33] Bai, H.; Mi, C. Eliminate Reactive Power and Increase System Efficiency of Isolated Bidirectional Dual-Active-Bridge DC–DC Converters Using Novel Dual-Phase-Shift Control. IEEE Trans. Power Electron. 2008, 23, 2905–2914.

[34] Jain, A.K.; Ayyanar, R. Pwm control of dual active bridge: Comprehensive analysis and experimental verification. IEEE Trans. Power Electron. 2010, 26, 1215–1227.

[35] Jauch, F.; Biela, J. Modelling and ZVS control of an isolated three-phase bidirectional AC–DC converter. In Proceedings of the 2013 15th European Conference on Power Electronics and Applications (EPE), Lille, France, 2–6 September 2013; pp. 1–11.

[36] Friedli, T.; Kolar, J.W.; Rodriguez, J.; Wheeler, P.W. Comparative Evaluation of Three-Phase AC–AC Matrix Converter and Voltage DC-Link Back-to-Back Converter Systems. IEEE Trans. Ind. Electron. 2011, 59, 4487–4510.

[37] L. Huber, J. Yungtaek, and M. M. Jovanovic, "Performance evaluation of bridgeless PFC boost rectifiers," IEEE Trans. Power Electronics, vol. 23, no. 3, pp. 1381-1390, May 2008.

[38] F. Musavi, W. Eberle, and W. G. Dunford, "A high-performance single-phase bridgeless interleaved PFC converter for plug-in hybrid electric vehicle battery chargers," IEEE Trans. Industry Applications, vol. 47, no. 4, pp. 1833-1843, Aug. 2011.

[39] O. Chang-Yeol, K. Dong-Hee, W. Dong-Gyun, S. Won-Yong, K. Yun-Sung, and L. Byoung-Kuk, "A high-efficient nonisolated single-stage on-board battery charger for electric vehicles," IEEE Trans. Power Electronics, vol. 28, no. 12, pp. 5746-5757, Dec. 2013.

[40] R. M. Schupbach and J. C. Balda, "Comparing DC–DC converters for power management in hybrid electric vehicles," IEEE Int. Electric Machines and Drives Conference (IEMDC), Madison, WI, Jun. 1-4, 2003, pp. 1369-1374.
[41] V. Kaura and V. Blasko, "Operation of a phase locked loop system under distorted utility conditions," IEEE Trans. Industry Applications, vol. 33, no. 1, pp. 58-63, Jan./Feb. 1997.
[42] F. F. Judd and C. T. Chen, "On the performance and design of self-oscillating DC-to-DC converters," IEEE Trans. Industrial Electronics and Control Instrumentation, vol. 19, no. 3, pp. 89-97, Aug. 1972.
[43] Z. Lai and K. M. Smedley, "A family of continuous-conduction-mode power-factor correction controllers based on the general pulse-width modulator," IEEE Trans. Power Electronics, vol. 13, no. 3, pp. 501-510, May 1998.
[44] L. Roggia, F. Beltrame, J. E. Baggio, and J. R. Pinheiro, "Digital current controllers applied to the boost power factor correction converter with load variation," IET Power Electronics, vol. 5, no. 5, pp. 532-541, May 2012.
[45] Newman, J.; Thomas, K.E.; Hafezi, H.; Wheeler, D.R. Modeling of lithium-ion batteries. J. Power Sources 2003, 119–121, 838–843.
[46] Vertiz, G.; Oyarbide, M.; Macicior, H.; Miguel, O.; Cantero, I.; de Arroiabe, P.F.; Ulacia, I. Thermal characterization of large size lithium-ion pouch cell based on 1d electro-thermal model. J. Power Sources 2014, 272, 476–484.
[47] Park, S.; Jung, D. Battery cell arrangement and heat transfer fluid effects on the parasitic power consumption and the cell temperature distribution in a hybrid electric vehicle. J. Power Sources 2013, 227, 191–198.
[48] Park, H. A design of air flow configuration for cooling lithium ion battery in hybrid electric vehicles. J. Power Sources 2013, 239, 30–36.
[49] Tong, W.; Somasundaram, K.; Birgersson, E.; Mujumdar, A.S.; Yap, C. Numerical investigation of water cooling for a lithium-ion bipolar battery pack. Int. J. Therm. Sci. 2015, 94, 259–269.
[50] Zhao, J.; Rao, Z.; Liu, C.; Li, Y. Experimental investigation on thermal performance of phase change material coupled with closed-loop oscillating heat pipe (PCM/CLOHP) used in thermal management. Appl. Therm. Eng. 2016, 93, 90–100.
[51] Tran, T.-H.; Harmand, S.; Desmet, B.; Filangi, S. Experimental investigation on the feasibility of heat pipe cooling for HEV/EV lithium-ion battery. Appl. Therm. Eng. 2014, 63, 551–558.

[52] Burger, S.; Chaves-Ávila, J.P.; Batlle, C.; Pérez-Arriaga, I.J. A Review of the Value of Aggregators in Electricity Systems. Renew.Sustain. Energy Rev. 2017, 77, 395–405.
[53] Gkatzikis, L.; Koutsopoulos, I.; Salonidis, T. The Role of Aggregators in Smart Grid Demand Response Markets. IEEE J. Sel. Areas Commun. 2013, 31, 1247–1257.
[54] Lin, H.; Liu, Y.; Sun, Q.; Xiong, R.; Li, H.; Wennersten, R. The Impact of Electric Vehicle Penetration and Charging Patterns on the Management of Energy Hub–A Multi-Agent System Simulation. Appl. Energy 2018, 230, 189–206.
[55] Yu, R.; Ding, J.; Zhong, W.; Liu, Y.; Xie, S. PHEV Charging and Discharging Cooperation in V2G Networks: A Coalition Game Approach. IEEE Internet Things J. 2014, 1, 578–589.
[56] Luo, C.; Huang, Y.-F.; Gupta, V. Stochastic Dynamic Pricing for EV Charging Stations With Renewable Integration and Energy Storage. IEEE Trans. Smart Grid 2017, 9, 1494–1505.
[57] Ansari, M.; Al-Awami, A.T.; Sortomme, E.; Abido, M.A. Coordinated bidding of ancillary services for vehicle-to-grid using fuzzy optimization. IEEE Trans. Smart Grid 2014, 6, 261–270.
[58] Buja, G.; Bertoluzzo, M.; Fontana, C. Reactive Power Compensation Capabilities of V2G-Enabled Electric Vehicles. IEEE Trans. Power Electron. 2017, 32, 9447–9459.

3

Wireless Charging Schemes for EVs

Khalid Yahya[1], Mohamed Salem[2], Adel E. M. Yahya[3], Nassim A. Iqteit[4], and N. Prabaharan[5]

[1]Department of Electrical and Electronics Engineering, Faculty of Engineering and Architecture, Istanbul Gelisim University, Istanbul 34310, Turkey
[2]School of Electrical and Electronic Engineering, Universiti Sains Malaysia (USM), Malaysia
[3]Department of Electrical Engineering, Regdalen, Sabratah University, Libya
[4]Department of Electrical Engineering, Palestine Polytechnic University, Palestine
[5]Department of Electrical & Electronics Engineering, SASTRA Deemed University, India
E-mail: hayahya@gelisim.edu.tr; salemm@usm.my; ecc.yahya@gmail.com; nassim_eng83@ppu.edu; prabaharan.nataraj@gmail.com

Abstract

Wireless charging technology has emerged as a promising solution for electric vehicle (EV) charging. This chapter provides an overview of wireless charging schemes for EVs and discusses the latest advancements in the field. The first section of the chapter presents an introduction to wireless charging technology for EVs and compares it to traditional plug-in charging. The advantages and limitations of wireless charging are also discussed.

The chapter then delves into magnetic resonance-based wireless charging, exploring the concept and reviewing the latest advancements in the technology. The potential of magnetic resonance-based wireless charging for EVs is analyzed. The inductive wireless charging schemes and topologies are also discussed, along with the performance, efficiency, and safety of inductive charging for EVs.

The concept of dynamic wireless charging is introduced in the following section, along with its potential applications and technical challenges. A comparative analysis of different wireless charging schemes is presented, which includes an evaluation of their advantages and disadvantages. Factors that affect the selection of a suitable wireless charging scheme for EVs are also discussed.

The chapter concludes by summarizing the key findings and insights, as well as identifying the future directions and research opportunities for wireless charging technology for EVs. Overall, this chapter aims to provide an in-depth understanding of the latest wireless charging schemes for EVs, and to highlight the potential of this technology in revolutionizing the EV industry.

Keywords: Wireless charging, electric vehicles, magnetic resonance, inductive charging, dynamic wireless charging.

3.1 Introduction

The demand for electric vehicles (EVs) is growing exponentially, driven by the need to reduce greenhouse gas emissions, improve energy security, and enhance energy efficiency. The global EV market is projected to reach 230 million units by 2030, and many countries and companies are setting ambitious targets for electrifying their transportation sectors [1]. As the market for EVs grows, the need for convenient and efficient charging solutions becomes increasingly important. One promising solution is wireless charging technology, which eliminates the need for physical cables and plugs, and provides a more convenient and flexible way to charge EVs.

This chapter provides an overview of the latest wireless charging schemes for EVs, discussing their technical principles, performance, efficiency, safety, and practical applications. The chapter compares the different wireless charging schemes, highlighting their advantages and disadvantages, and provides insights into the future direction of wireless charging technology for EVs.

The chapter begins with an overview of wireless charging technology for EVs, describing the technical principles and how it differs from traditional plug-in charging. The comparison of wireless charging vs. traditional plug-in charging highlights the advantages and limitations of wireless charging. Wireless charging offers several advantages, including greater convenience, safety, and reduced wear and tear on the charging infrastructure. Moreover, wireless charging has the potential to facilitate autonomous charging, where the charging process is initiated and managed by the EV itself, allowing for a more seamless and automated charging experience. However, wireless charging efficiency and power transfer rates are still lower than those of plug-in charging, which can affect the range and charging time of EVs.

Despite the potential benefits, wireless charging for EVs is still a relatively new and developing technology, with several technical and practical challenges that need to be addressed. For example, wireless charging requires a certain level of electromagnetic compatibility (EMC) to prevent electromagnetic interference (EMI) with other electronic devices, and to ensure safety for users. This chapter provides insights into the latest research on addressing these technical challenges, and how they can be overcome to achieve greater efficiency, safety, and performance in wireless charging for EVs.

The purpose of this chapter is to provide a valuable resource for researchers, engineers, and EV manufacturers interested in wireless charging technology and its potential applications in the EV industry. The chapter draws on recent research studies and publications [2–5] to provide an up-to-date and comprehensive analysis of the subject.

3.2 Magnetic Resonance-based Wireless Charging

Wireless charging is a promising technology for electric vehicles (EVs) that eliminates the need for cables and connectors, simplifying the charging process for drivers. Magnetic resonance-based wireless charging (MRWC) is a type of wireless charging technology that uses resonant magnetic fields to transfer power from the charging infrastructure to the EV's battery.

Exploring the concept of magnetic resonance-based wireless charging, MRWC systems typically operate at a frequency of 85 kHz, and the charging pad and receiver coil are designed to resonate at this frequency. This resonance allows for efficient power transfer between the charging pad and the EV. The efficiency of an MRWC system is typically in the range of 90%–95%, which is comparable to that of traditional plug-in charging [6].

Figure 3.1 shows a schematic diagram of a typical MRWC system. The system consists of a charging pad, which is typically installed on the ground, and an EV with a receiver coil. The charging pad generates a resonant magnetic field, which induces a current in the receiver coil of the EV. This current charges the EV's battery.

One of the key advantages of MRWC is that it does not require a physical connection between the charging infrastructure and the EV. This eliminates the need for cables and connectors, which can be inconvenient and prone to

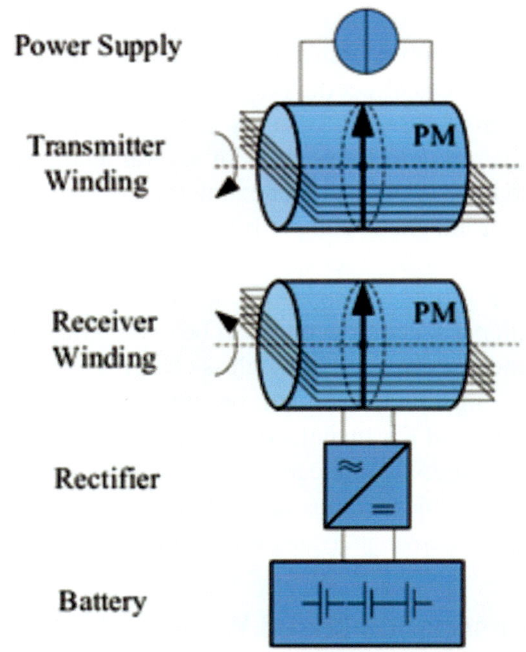

Figure 3.1 Schematic diagram of a typical MRWC system [7].

wear and tear over time. Additionally, since MRWC operates at a relatively low frequency, it does not pose any significant health risks to humans or animals [8].

The fundamental wireless power transfer (WPT) system based on coupled magnetic resonance consists of four power stages: the power factor correction (PFC) converter, the radio frequency (RF) amplifier, the resonators or coils, and the onboard rectifier. A schematic of a four-coil magnetic resonance wireless EV charging system with uncontrolled pick-up is shown in Figure 3.2. To achieve an overall efficiency of 90%, the PFC and RF amplifier stages must have a minimum efficiency of 97%, while the coil-to-coil efficiency must be higher than 96%, and the rectifier efficiency must be close to 99% [7].

The most crucial component of the wireless charging system is the design of the coil, which has a significant impact on the power capacity and efficiency of the system. The size of the coil determines the upper limit of the power capacity, while the quality factor of the coil affects its efficiency. In the case of vehicle applications, the coil comprises two components – the

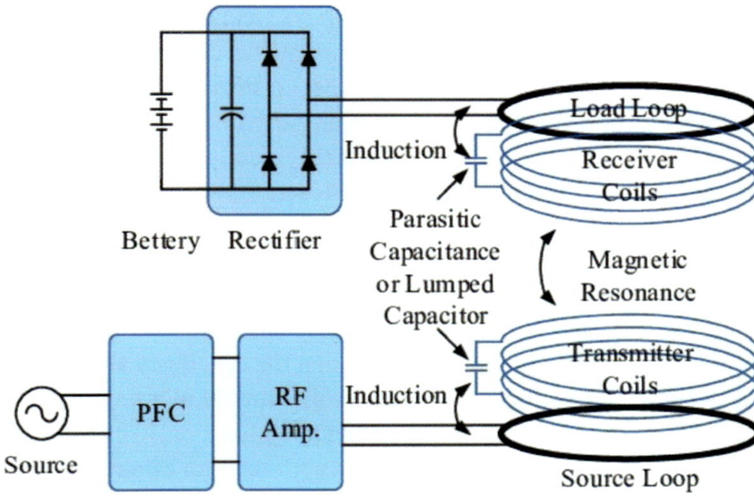

Figure 3.2 Schematic diagram of a typical MRWC system [7].

magnetics and wires – and operates within a frequency range of 10 kHz to several hundred Hertz. Due to the high frequency, the skin effect causes a high AC resistance, leading to the use of Litz wire [9]. The alternating current that flows through one strand induces an alternating magnetic field that creates eddy currents in the adjacent strands. This proximity effect also increases the AC resistance. In contrast, when operating under MHz, a solid copper wire may offer higher efficiency than Litz wire [10], as the proximity effect can dominate the AC resistance under such a frequency. To eliminate this effect, magnetic thin films coated on the wire can provide shielding for the alternating magnetic flux and prevent the eddy current [11]. Alternatively, nonmetallic materials like graphene and carbon nanotubes have been proposed to eliminate both the skin effect and proximity effect [12].

One of the most important equations that describes the transfer of energy through resonant magnetic coupling is the mutual inductance equation (eqn (3.1)), which can be expressed as:

$$M = k \times sqrt(L1 \times L2), \tag{3.1}$$

where M is the mutual inductance, $L1$ and $L2$ are the inductances of the two resonant coils, and k is the coupling coefficient between the coils. This equation is important because it helps to determine the strength of the coupling between the transmitting and receiving coils, which can affect the efficiency of the wireless charging system.

However, there are some limitations to MRWC technology. For example, the system is sensitive to misalignment between the charging pad and the receiver coil. If the EV is not parked in the correct position, the power transfer efficiency can be reduced, and the charging time can be significantly increased. Additionally, MRWC systems are typically more expensive to install than traditional plug-in charging infrastructure, which can be a barrier to adoption [13].

Reviewing the latest advancements in magnetic resonance-based charging technology, despite these limitations, MRWC technology has the potential to play a significant role in the future of EV charging. As research and development in this field continue, it is likely that the efficiency and reliability of MRWC systems will continue to improve, making them a more attractive option for EV owners and infrastructure providers alike.

While MRWC is a promising technology with its high efficiency and the advantage of no physical connection between the charging pad and the EV, it still faces some challenges like misalignment sensitivity and high installation costs. In contrast, inductive wireless charging operates at a higher frequency and relies on a direct physical connection between the charging pad and the EV. Let's explore more about inductive wireless charging technology and its benefits.

3.3 Inductive Wireless Charging

Inductive wireless charging, also known as wireless power transfer (WPT), is a popular technology for charging electric vehicles (EVs) that allows for contactless energy transfer between the charging infrastructure and the EV's battery. In this method, power is transferred through magnetic fields between two coils that are in close proximity to each other, one in the charging pad and the other in the EV. The coils are designed to resonate at the same frequency, creating a strong magnetic field that can transfer energy over a short distance.

3.3.1 Inductive charging schemes and topologies

There are two common inductive charging schemes used in the EV industry: stationary wireless charging and dynamic wireless charging. Stationary wireless charging refers to the process of charging a stationary EV without any movement, while dynamic wireless charging is the process of charging a moving EV. Dynamic wireless charging can be further divided into two topologies: under-road charging and on-road charging.

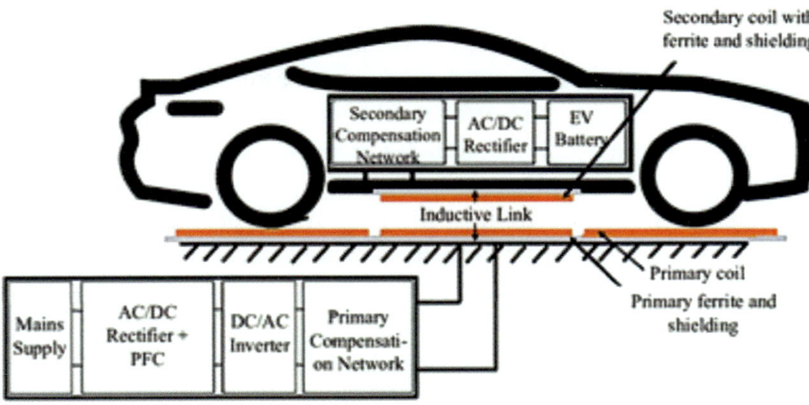

Figure 3.3 General structure of an inductive wireless charging system for EVs [14].

Under-road charging involves installing charging coils under the road surface, which are activated when the EV passes over them. On the other hand, on-road charging involves installing charging coils on the road surface, which are activated when the EV is parked over them or passes by at a low speed.

Figure 3.3 shows the general structure of an inductive wireless charging system for EVs, including the charging pad, receiver coil, rectifier, and battery.

3.3.1.1 Performance analysis of inductive charging for EVs

Inductive wireless charging technology offers several advantages over traditional plug-in charging methods. First, it eliminates the need for physical cables and connectors, making charging more convenient for EV owners. Second, it reduces the risk of electric shock or other safety hazards associated with plug-in charging. Finally, it can be used in various applications, including stationary charging and dynamic charging for EVs in motion.

However, inductive charging also has some limitations, including lower charging efficiency compared to plug-in charging, as well as sensitivity to the misalignment between the charging pad and the receiver coil, which can significantly reduce the charging efficiency.

The efficiency of the inductive charging system is determined by several factors, including the quality factor of the coils, the coupling coefficient, and the operating frequency. A higher quality factor and coupling coefficient can increase the system's efficiency, while a lower operating frequency can result in lower power loss.

Overall, inductive wireless charging is a promising technology for the EV industry, with the potential to improve the convenience and safety of EV charging. Ongoing research and development in this field will continue to enhance the performance and efficiency of inductive charging systems, making them a more attractive option for EV owners and infrastructure providers.

In conclusion, inductive wireless charging is a promising technology that offers many advantages for electric vehicle owners. It simplifies the charging process by eliminating the need for cables and connectors, and it poses no significant health risks to humans or animals. While there are some limitations to inductive wireless charging, research and development in this field are ongoing, and it is likely that the technology will continue to improve in the coming years. As the market for electric vehicles continues to grow, inductive wireless charging has the potential to play an increasingly important role in the future of electric vehicle charging infrastructure.

Moving on to capacitive wireless charging, this technology offers an alternative approach to wireless charging that is based on the principle of capacitance. We will explore the concept of capacitive wireless charging, including its key advantages and limitations, as well as the most common topologies and performance metrics.

3.3.2 Capacitive wireless charging for EVs

Capacitive wireless charging (CWC) is another type of wireless charging technology that uses electric fields to transfer power from the charging infrastructure to the EV's battery. In contrast to inductive wireless charging, which uses magnetic fields, CWC technology relies on electric fields to induce a current in the receiver coil of the EV.

One of the key advantages of CWC technology is that it is less sensitive to misalignment between the charging pad and the receiver coil compared to inductive wireless charging. This is because electric fields are less affected by the distance between the charging pad and the receiver coil, which means that the charging pad and the EV can be further apart without significant loss of efficiency [15].

Figure 3.4 shows a schematic diagram of a typical CWC system. The system consists of a charging pad, which is typically installed on the ground, and an EV with a receiver plate. The charging pad generates an electric field, which induces a current in the receiver plate of the EV. This current charges the EV's battery.

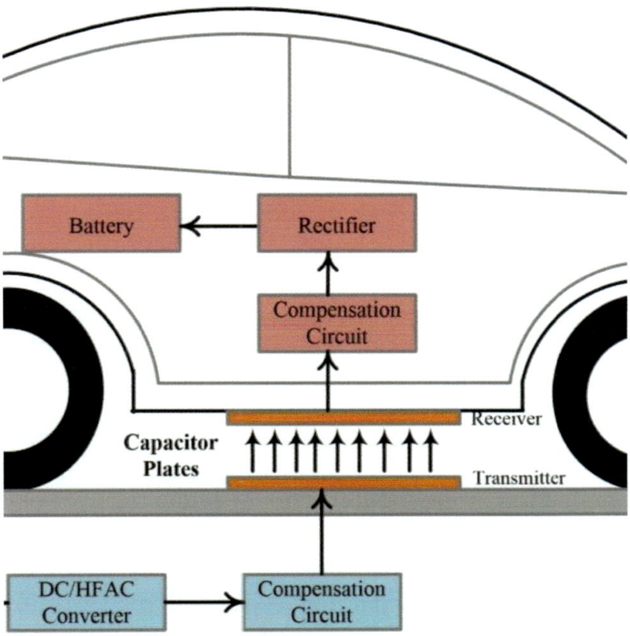

Figure 3.4 Schematic diagram of a typical CWC system.

Like MRWC and inductive wireless charging, the efficiency of CWC depends on various factors such as the distance between the charging pad and the receiver plate, the size and shape of the charging pad, and the efficiency of the power electronics used in the system. One of the key challenges in CWC technology is to maintain a high charging efficiency while keeping the electric fields generated by the charging pad within safe limits for humans and animals. This requires careful design of the charging pad and the power electronics used in the system.

The efficiency of CWC is typically lower than that of MRWC and inductive wireless charging, with typical values in the range of 70%–80%. However, recent research has shown that by optimizing the design of the charging pad and the power electronics, it is possible to achieve efficiencies of up to 90% [7].

Figure 3.5 shows the efficiency of a CWC system as a function of the distance between the charging pad and the receiver plate. As can be seen from the figure, the efficiency of the system drops rapidly as the distance between the charging pad and the receiver plate increases. This is because the electric fields generated by the charging pad weaken as they propagate through the

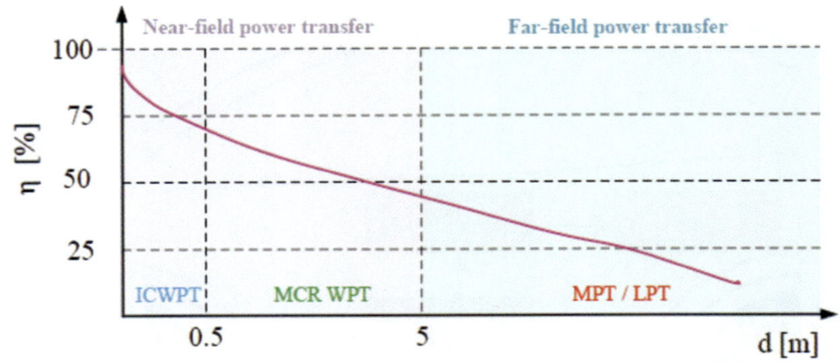

Figure 3.5 Efficiency of the system as a function of the distance between the charging pad and the receiver plate [16].

air, leading to a reduction in the induced current in the receiver plate of the EV.

Despite the challenges, CWC technology has the potential to play an important role in the future of EV charging, especially in applications where the distance between the charging pad and the receiver plate is relatively large. As research and development in this field continue, it is likely that the efficiency and reliability of CWC systems will continue to improve, making them a more attractive option for EV owners and infrastructure providers alike.

3.3.2.1 Performance analysis of capacitive wireless charging for EVs

Capacitive wireless charging is a relatively new technology that has the potential to revolutionize the way electric vehicles are charged. However, like all new technologies, it is important to consider the performance, efficiency, and safety of the system before widespread adoption.

One of the main advantages of capacitive wireless charging is its ability to operate at higher frequencies than inductive charging, which can result in higher power transfer rates. Capacitive wireless charging systems typically operate in the range of 6.78–13.56 MHz, with power transfer rates ranging from a few hundred watts to several kilowatts [17].

The efficiency of a capacitive wireless charging system is determined by several factors, including the distance between the charging pad and the receiver coil, the size and shape of the capacitive plates, and the dielectric constant of the material between the plates. In general, capacitive wireless charging systems have lower efficiency than inductive systems, with typical efficiencies in the range of 70%–80% [18]. However, recent research has shown that efficiency can be improved through careful design of the charging pad and receiver coil, as well as the use of high-permittivity dielectric materials.

Safety is also a critical consideration for any wireless charging system. Capacitive wireless charging systems have the advantage of operating at higher frequencies than inductive systems, which reduces the risk of electromagnetic interference with sensitive electronics in the vehicle. However, there are also concerns about the potential for electric shock or burns if a person comes into contact with the charging pad or receiver coil. Safety features such as automatic shutdown and grounding of the charging pad can mitigate these risks.

Overall, capacitive wireless charging shows great promise as a fast and efficient method for charging electric vehicles. While there are still some challenges to overcome, ongoing research and development in this field is expected to lead to further improvements in performance, efficiency, and safety of capacitive wireless charging systems for EVs.

3.3.3 Dynamic wireless charging for EVs

Dynamic wireless charging (DWC) is a type of wireless charging technology that allows electric vehicles to charge while they are in motion. This technology uses an electromagnetic field to transfer power from the roadway to the vehicle. The roadway is equipped with a series of coils that generate a magnetic field, which is picked up by a receiver coil on the underside of the vehicle. The magnetic field induces an electric current in the receiver coil, which charges the vehicle's battery.

One of the advantages of dynamic wireless charging is that it can provide continuous charging while the vehicle is in motion, eliminating the need for frequent stops to recharge the battery. This can be particularly useful for electric vehicles with large batteries, which may require long charging times. Additionally, dynamic wireless charging can reduce the size and weight of the battery required for an electric vehicle, which can improve its efficiency and performance.

Figure 3.6 shows a schematic diagram of a dynamic wireless charging system. The system consists of a series of coils embedded in the roadway, which generate a magnetic field, and a receiver coil on the underside of the vehicle, which picks up the magnetic field and converts it into an electric

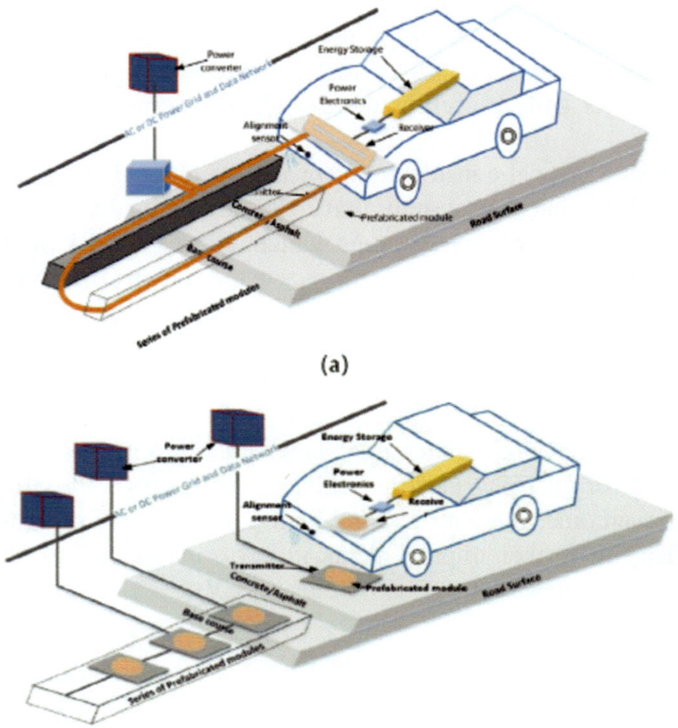

Figure 3.6 Schematic diagram of a dynamic wireless charging system [19].

current. The efficiency of a dynamic wireless charging system depends on factors such as the distance between the roadway coils and the receiver coil, the speed of the vehicle, and the power output of the charging system.

Several research projects and pilot programs have been conducted to evaluate the feasibility and performance of dynamic wireless charging technology. One of the largest and most well-known projects is the German-funded "eHighway" program, which tested dynamic wireless charging technology on a 10-kilometer stretch of highway in Germany [20]. The program demonstrated that dynamic wireless charging is a viable technology that can significantly extend the range of electric vehicles.

However, there are some challenges associated with dynamic wireless charging technology. One of the primary challenges is the cost and complexity of installing the charging infrastructure on roadways. The roadway must be excavated to install the charging coils, which can be time-consuming and expensive.

Despite these challenges, dynamic wireless charging has the potential to revolutionize the way electric vehicles are charged and used. As research and development in this field continue, it is likely that the efficiency, reliability, and cost-effectiveness of dynamic wireless charging systems will continue to improve, making them a more practical and attractive option for electric vehicle owners and infrastructure providers alike.

3.3.3.1 Performance analysis of dynamic wireless charging for EVs

Dynamic wireless charging is a technology that allows EVs to be charged while in motion, without the need for stops at charging stations. This technology uses a series of charging pads installed on the road surface to transfer power to the EV's battery as it drives over them. This technology has the potential to increase the range and flexibility of EVs, as they can be charged while in use, which could be particularly useful for public transportation and other fleet vehicles.

The efficiency of a DWC system is affected by several factors, including the speed of the vehicle, the size and number of the charging pads, and the distance between the charging pads and the EV's receiver coil. As with other wireless charging technologies, DWC systems typically operate at high frequencies (in the range of several megahertz), which allows for efficient power transfer between the charging pads and the EV.

Studies have shown that DWC systems can achieve transfer efficiencies of up to 90% [7]. However, it should be noted that the efficiency of DWC systems can be affected by several factors, such as the distance between the charging pads and the receiver coil of the EV. The charging pads must be precisely placed, and the EV must be equipped with the appropriate receiver coil to ensure efficient power transfer.

One of the key advantages of DWC is its potential to provide uninterrupted charging for EVs, eliminating the need for regular stops at charging stations. This can be particularly useful for public transportation systems, such as buses and trams, which can be charged while in operation. Additionally, since DWC operates at high frequencies, it does not pose any significant health risks to humans or animals.

However, there are some limitations to DWC technology. For example, the cost of installing a DWC system can be prohibitively high, particularly for large-scale implementation. Additionally, since the system is installed on the road surface, there may be concerns about the durability and lifespan of the charging pads, particularly in areas with heavy traffic.

Overall, DWC technology has the potential to revolutionize the way we charge EVs, allowing for uninterrupted charging while in use. As research and development in this field continue, it is likely that the efficiency and reliability of DWC systems will continue to improve, making them a more attractive option for EV owners and infrastructure providers alike.

3.4 Wireless Charging Roadway Systems

One of the most common wireless charging roadway schemes is known as inductive power transfer (IPT). In an IPT system, the charging infrastructure is embedded in the roadway, while the EV is equipped with a receiver coil. As the EV travels over the charging infrastructure, a resonant magnetic field is generated, which induces a current in the receiver coil, charging the EV's battery.

Figure 3.7 shows a schematic diagram of an IPT-based wireless charging roadway system. The system consists of a charging infrastructure, which is typically embedded in the roadway, and an EV with a receiver coil. As the EV drives over the charging infrastructure, a resonant magnetic field is generated, which induces a current in the receiver coil of the EV, charging the battery.

Another wireless charging roadway scheme is known as dynamic wireless charging (DWC). In a DWC system, the charging infrastructure is installed in the roadway, and the EV is equipped with a receiver coil and a battery. As

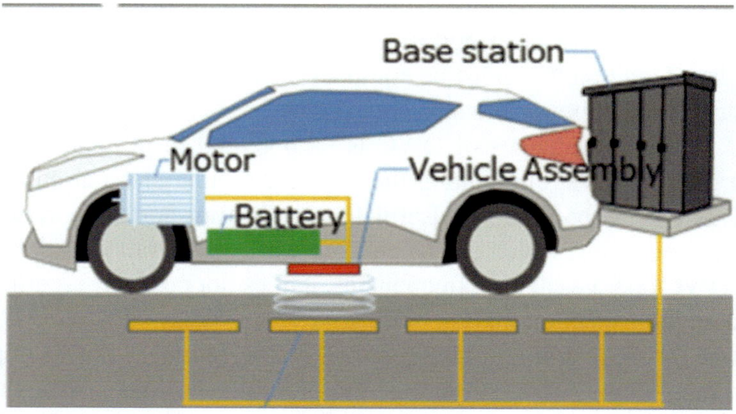

Figure 3.7 Schematic diagram of an inductive power transfer-based wireless charging roadway system [21].

the EV drives over the charging infrastructure, power is wirelessly transferred from the roadway to the EV, charging the battery.

One of the key advantages of wireless charging roadway systems is that they provide a seamless and continuous charging experience for EV drivers. Instead of having to stop and charge at individual charging stations, drivers can simply drive on a roadway with embedded charging infrastructure and have their EVs charged wirelessly. This can significantly reduce the time and inconvenience associated with traditional charging methods.

However, there are also some limitations to wireless charging roadway systems. For example, the cost of installing the charging infrastructure in the roadway can be significant, which can be a barrier to adoption. Additionally, since the charging infrastructure is embedded in the roadway, maintenance and repairs can be challenging and expensive.

3.4.1 Performance analysis of wireless charging roadway systems for EVs

In terms of performance, efficiency, and safety, wireless charging roadway systems are still being developed and optimized. However, research has shown that these systems can achieve high power transfer efficiencies, in the range of 90%. Safety is also a concern, as the charging infrastructure is embedded in the roadway, which can pose risks to pedestrians and animals. Therefore, strict safety standards and guidelines must be followed to ensure the safe deployment of wireless charging roadway systems.

Overall, wireless charging roadway systems have the potential to play a significant role in the future of EV charging infrastructure. As research and development in this field continue, it is likely that the efficiency and reliability of these systems will continue to improve, making them a more attractive option for EV owners and infrastructure providers alike.

3.5 Comparative Analysis of Wireless Charging Schemes

Table 3.1 provides a comparative analysis of different wireless charging schemes for electric vehicles (EVs). The table presents key factors such as charging time, advantages, disadvantages, efficiency, and safety of each wireless charging scheme. This analysis can help EV owners and infrastructure providers choose the most suitable charging solution for their needs.

Table 3.1 Different wireless charging schemes for electric vehicles (EVs).

Charging scheme	Charging time (hours)	Advantages	Disadvantages	Efficiency	Safety
Inductive	2–6	No physical contact, relatively high efficiency	Sensitive to misalignment, more expensive installation	90%–95%	Low health risks at low frequency
Capacitive	2–6	No physical contact, simple topology	Limited range, low efficiency	70%–80%	Potential for electric shock
Dynamic	Continuous	Continuous charging, no need for charging stops	Expensive infrastructure, low efficiency	70%–75%	Safety concerns with exposed conductive roadway
Roadway	Continuous	Continuous charging, no need for charging stops, less battery weight	Expensive infrastructure, limited applicability	Up to 95%	Safe for humans and animals

Sure, I can include the roadway scheme in the table. Here's an updated version of the table with the roadway scheme and the requested columns:

Overall, each wireless charging scheme has its own advantages and disadvantages. Inductive charging is currently the most widely used and tested wireless charging technology, but it is limited by its sensitivity to misalignment and high installation cost. Capacitive charging is a simpler and potentially more cost-effective option, but it is also limited by its power transfer capabilities and the need for a larger gap between the charging pad and the vehicle. Dynamic and roadway charging have the potential to revolutionize the EV industry by eliminating range anxiety, but both require extensive infrastructure and road modifications and are currently limited in their compatibility with existing EVs. As research and development continue in this field, it is likely that the efficiency, and safety of wireless charging schemes will improve, making them a more attractive option for both EV owners and infrastructure providers.

3.6 Conclusion and Future Directions

In conclusion, this chapter has provided an overview of the different wireless charging technologies for electric vehicles, including inductive, capacitive, and dynamic wireless charging, as well as the emerging technology of wireless charging roadway systems. A comparative analysis of these different technologies has been presented, highlighting the advantages, disadvantages, efficiency, and safety of each.

Overall, it is clear that wireless charging technology has the potential to revolutionize the way we charge electric vehicles, making the process more convenient and user-friendly. However, there are still some limitations and challenges that must be overcome, including cost, efficiency, safety, and standardization.

Based on the discussion presented in this chapter, we can highlight several key points:

1. Wireless charging technology has the potential to significantly simplify the charging process for electric vehicles, making EVs more practical and convenient for drivers.
2. There are multiple wireless charging schemes and topologies, each with its own advantages and disadvantages. The most common wireless charging schemes are magnetic resonance-based, capacitive, and dynamic wireless charging.
3. Magnetic resonance-based wireless charging is the most mature wireless charging technology and is already being used in some EVs. Capacitive and dynamic wireless charging are still in the early stages of development and are not yet widely used.
4. The efficiency and safety of wireless charging systems are key factors that must be considered in the design and implementation of these systems.
5. Wireless charging roadway systems have the potential to revolutionize the EV industry by enabling EVs to charge while driving. However, the practical challenges of implementing these systems, such as ensuring safety and maximizing efficiency, must be carefully addressed.

In the future, continued research and development in wireless charging technology will be necessary to improve the efficiency and reliability of these systems. As wireless charging technology becomes more mature and cost-effective, it is likely that we will see wider adoption of this technology in the EV industry. Additionally, the integration of wireless charging technology into smart grid systems and the development of new business models for EV charging will be important areas for future research and development.

Overall, the future of wireless charging for electric vehicles looks bright, and we are excited to see what new and innovative technologies will emerge in the years to come.

References

[1] International Energy Agency, "Global EV Outlook 2021: Scaling up the transition to electric mobility," 2021. [Online]. Available: https://www.iea.org/reports/global-ev-outlook-2021. [Accessed: Feb. 17, 2023].

[2] Alhamrouni, Ibrahim, M. Iskandar, Mohamed Salem, Lilik J. Awalin, Awang Jusoh, and Tole Sutikno. "Application of inductive coupling for wireless power transfer." International Journal of Power Electronics and Drive Systems 11, no. 3 (2020): 1109.

[3] D. H. Nguyen, "Dynamic Optical Wireless Power Transfer for Electric Vehicles," in IEEE Access, vol. 11, pp. 2787-2795, 2023, doi:10.1109/ACCESS.2023.3234577.

[4] F. Grazian, T. B. Soeiro and P. Bauer, "Inductive Power Transfer Based on Variable Compensation Capacitance to Achieve an EV Charging Profile With Constant Optimum Load," in IEEE Journal of Emerging and Selected Topics in Power Electronics, vol. 11, no. 1, pp. 1230-1244, Feb. 2023, doi:10.1109/JESTPE.2022.3188060.

[5] Y. Yang, J. Wang, Z. Huang, I. -W. Iam and C. -S. Lam, "Automatic Containment of Field Exposure for Roadway Wireless Electric Vehicle Charger," in IEEE Transactions on Transportation Electrification, doi: 10.1109/TTE.2023.3236684.

[6] H. H. Wu, A. Gilchrist, K. Sealy, P. Israelsen and J. Muhs, "A review on inductive charging for electric vehicles," 2011 IEEE International Electric Machines & Drives Conference (IEMDC), Niagara Falls, ON, Canada, 2011, pp. 143-147, doi:10.1109/IEMDC.2011.5994820.

[7] C. Qiu, K. T. Chau, C. Liu and C. C. Chan, "Overview of wireless power transfer for electric vehicle charging," 2013 World Electric Vehicle Symposium and Exhibition (EVS27), Barcelona, Spain, 2013, pp. 1-9, doi:10.1109/EVS.2013.6914731.

[8] A. El-Shahat, J. Danjuma, A. Y. Abdelaziz, and S. H. E. Abdel Aleem, "Human Exposure Influence Analysis for Wireless Electric Vehicle Battery Charging," Clean Technologies, vol. 4, no. 3, pp. 785-805, 2022. DOI:10.3390/cleantechnol4030048.

[9] J.M. Miller, C.P. White, O.C. Onar, and P.M. Ryan, "Grid side regulation of wireless power charging of plug-in electric vehicles," in Proc. Energy Conv. Congr. Expo. Raleigh, NC, 2012, pp. 261-268.
[10] WiTricity, WiT-3300, [Online]. Available: http://www.witricity.com/pages/evcharging-system.html.
[11] S. Lee, J. Huh, C. Park, N.S. Choi, G.H. Cho, and C.T. Rim, "On-Line Electric Vehicle using inductive power transfer system," in Proc. Energy Convers. Congr. Expo. Atlanta, GA, 2010, pp. 1598-1601.
[12] S. Ahn, N.P. Suh, and D.H. Cho, "Charging up the road," IEEE Spectr. Vol.50, no. 4, pp. 48-54, Apr. 2013.
[13] M. R. Khalid, I. A. Khan, S. Hameed, M. S. J. Asghar and J. -S. Ro, "A Comprehensive Review on Structural Topologies, Power Levels, Energy Storage Systems, and Standards for Electric Vehicle Charging Stations and Their Impacts on Grid," in IEEE Access, vol. 9, pp. 128069-128094, 2021, doi:10.1109/ACCESS.2021.3112189.
[14] E. ElGhanam, M. Hassan, and A. Osman, "Design of a High Power, LCC-Compensated, Dynamic, Wireless Electric Vehicle Charging System with Improved Misalignment Tolerance," Energies, vol. 14, no. 4, p. 885, Feb. 2021. doi: 10.3390/en14040885.
[15] Y. Gao, K. B. Farley, A. Ginart, and Z. T. H. Tse, "Safety and efficiency of the wireless charging of electric vehicles," Proceedings of the Institution of Mechanical Engineers, Part D: Journal of Automobile Engineering, vol. 230, no. 9, pp. 1196-1207, Aug. 2016. doi:10.1177/0954407015603863.
[16] K. Detka and K. Górecki, "Wireless Power Transfer—A Review," Energies, vol. 15, no. 19, p. 7236, Oct. 2022. doi:10.3390/en15197236.
[17] M. Z. Erel, K. C. Bayindir, M. T. Aydemir, S. K. Chaudhary and J. M. Guerrero, "A Comprehensive Review on Wireless Capacitive Power Transfer Technology: Fundamentals and Applications," in IEEE Access, vol. 10, pp. 3116-3143, 2022, doi: 10.1109/ACCESS.2021.3139761.
[18] D. Vincent, "Hybrid Inductive and Capacitive Wireless Power Transfer System for Future Transportation Electrification," Ph.D. dissertation, School of Graduate and Postdoctoral Studies, University of Ontario Institute of Technology, Oshawa, Ontario, Canada, Aug. 2020.
[19] C. Panchal, S. Stegen, and J. Lu, "Review of static and dynamic wireless electric vehicle charging system," Engineering Science and Technology, an International Journal, vol. 21, no. 5, pp. 922-937, Oct. 2018. doi: 10.1016/j.jestch.2018.06.015.

[20] Siemens AG, "eHighway–Solutions for electrified road freight transport," Nov. 8, 2017. [Online]. Available: https://press.siemens.com/global/en/fea.
[21] Toyota Motor Corporation, "Toyota's Humanoid Robot T-HR4 Makes a Leap Forward in Real-World Applications," Dec. 1, 2021. [Online]. Available:https://www.toyota-global.com/innovation/partner_robot/news/202112_01.html. [Accessed: Feb. 17, 2023].

4

Power Quality Assessment and Improvement of Grid-tied Distributed Generation Systems

M. Yu. Artemenko[1], V. V. Chopyk[2], V. M. Mykhalskyi[2], and I. A. Shapoval[2]

[1]Department of Acoustic and Multimedia Electronic Systems of the National Technical University of Ukraine, Igor Sikorsky Kyiv Polytechnic Institute, Ukraine
[2]Institute of Electrodynamics of the National Academy of Sciences of Ukraine, Ukraine
E-mail: artemenko_m_ju@ukr.net; chopykied@gmail.com; mikhalsky@ied.org.ua; shapoval@ied.org.ua

Abstract

The instantaneous power theory of multiphase power systems was extended by deriving new formulas for the instantaneous active current, instantaneous apparent power, and instantaneous power factor. These new formulas took into account the ratio of resistances between the different wires in the transmission line. The strategy of active filtering that forming this instantaneous active current in the transmission line prevails in the instantaneous power losses strategies based on the orthogonal decomposition of load currents using *pq*-theory of instantaneous power and its modification, cross-vector theory, *pqr*-theory of instantaneous power, proportional-vector theory with deprivation of the zero-sequence voltage component. The new formula for apparent power of the multiphase power system with resistive model of transmission line was proposed. It represents the geometric mean value calculated from the maximum permissible power loss along the transmission line and the short-circuit power of the multiphase source, with respect to the resistances in the transmission line. The formation of the active current vector by shunt active filter in the transmission line of a multiphase power supply system provides a minimum power loss and a unit value of the power factor. This filtering strategy outperforms other known filtering strategies based on active current formulas of authorship S. Fryze and P. Salmeron / J. Montano. The results of the comparative analysis show that the proposed formula for calculating the apparent power of a three-phase four-wire power system is equivalent to the formula specified in the IEEE standard and is fully consistent with the European approach to determining the apparent power in the case of sinusoidal power supply mode. It is established that for a given symmetrical voltage source the sinusoidal process of a three-phase four-wire power system with asymmetrical load is completely determined by six scalar quantities: active and reactive power, orthogonal components of unbalance power and orthogonal component of power associated with neutral line current. These quantities were used for active filtering control in grid-connected distributed generation systems. Integral formulas for determining the inactive power are derived to capture the contribution to power quality degradation. These formulas show that the output value of inactive power is preserved even in the presence of higher-order harmonic currents. The relationship between the unbalance power component and the linear parameters of the unbalanced load has been established, and computer experiments have verified these integral formulas.

Keywords: Apparent power, active current, active filtration, orthogonal components.

4.1 Introduction

Application of the great capabilities of modern control systems and power electronics to increase reliability and expand the functionality of power systems, led to the creation a new concept of electricity supply, known as distributed generation (DG) [1]. In the traditional concept of electricity supply, electric energy generation occurs in favorable conditions for its production, which are usually located far from the final consumers [2]. DG involves generating a significant amount of electricity using autonomous or integrated energy systems near consumption points where power interruption is possible. This difference in DG encourages renewable energy production and provides flexibility in the distribution of electricity following the needs of the load. Photovoltaic stations, wind turbines, fuel cells, microturbines, mobile electric vehicles are the primary energy resources of distributed generation systems. The compatibility of DG systems with traditional power supply systems is ensured by synchronous or asynchronous generators and, in recent years, mainly by power semiconductor converters [3].

Semiconductor converters with the possibility of bidirectional energy flow are crucial elements of microgrids and distributed generation systems [4, 5]. Different configurations of variable-speed wind turbines require semiconductor converters based on two inverters. Each inverter requires controlling active and reactive power to get optimal power from the wind turbine, exchanging reactive power with the electrical network. Similarly, photovoltaic systems need an inverter to inject the generated energy into the grid. Battery-based power storage systems also require bidirectional converters to charge the battery or supply power to the grid.

DG can work both offline and in network mode. DG's autonomous operation provides power to local loads, but there is no connection between the shared power network and the DG system. In addition, DG can operate in network connection mode when the generated power is fed into the network to power part of the load. The network connection mode of operation may contain several DG units that produce a significant amount of electric energy to meet the conditions of critical load to improve the system's reliability, primarily where there are violations in the supply of electricity. Power active filters are used in a distributed system connected to the network to integrate DG and eliminate harmonics in nonlinear loads.

In the traditional power supply system frames, power terms such as active power, reactive power, apparent power, and general harmonic distortions have been well-defined in sinusoidal operating conditions for linear and balanced symmetric multiphase or single-phase systems. However, due to the nonlinearity of loads and uncertainty in fast-growing distributed renewable generation sources within the new microgrid technology, active power balancing, reactive power supply, and harmonic compensation are becoming new critical issues [6]. Therefore, the lack of proper definitions for power components in the case of distorted current and voltage signals prompted the creation of new theories of power [7]. Active filtration is a compensation strategy that is widely implemented to counter all undesirable components that appear in the components of current and voltage [7]. The main functionality of all methods of shunt active filtration is the calculation of the compensating reference current, which is used in controllers to control the form of current. Standard power terms definitions in nonlinear and unbalanced systems need to be revised as new equations used in active filtration systems. This context requires a comprehensive review of power theories to be updated to calculate compensation currents in active filtration systems.

The use of multifunctional DG becomes the most important way to improve power quality in highly interactive systems, such as microgrids with dynamic load behaviour. The presence of reactive power and the circulation of asymmetric and harmonic currents can impair the overall operation of the equipment and the network itself [8]. In addition, the use of multifunctionality in decentralized generating plants is discussed in the literature, primarily based on the scenario of local application [9]. Since local multifunctional control usually does not consider the simultaneous operation of parallel inverters, its use can lead to undesirable interactions between different distributed generation systems. It can lead to inadequate compensations for current and voltage disturbances. To solve the problem of proper coordination of DG, the literature proposes control strategies to ensure the regulation of active and reactive power and the study of the provision of auxiliary services related to the quality of electric energy [5, 10].

DG coordination methodologies in the literature are mainly divided into three categories (centralized, decentralized, and distributed control), depending on how they relate to communication infrastructure [10]. Centralized methods, for example, rely on data channels to exchange information between inverters and the central control system. Distributed approaches exchange control information between participating parts (i.e., DG) that do not depend on a central controller, while decentralized inverter coordination does not typically use communication tools. Finally, the concept of hierarchical control [11] was proposed in the literature using each of the three categories mentioned above, implementing control systems based on a hierarchy of control levels.

Figure 4.1 shows an example of a microgrid connected to the main power grid. Such a microgrid feeds loads, including charging batteries of electric vehicles and contains power-generating elements in the form of a PVG photovoltaic generator, a WG wind turbine, and a battery storage element. A central controller controls the microgrid. The provision of bidirectional energy flows of battery storage and car charger and energy interfaces between renewable sources and the microgrid is carried out by semiconductor converters, shown in Figure 4.1 with its DC and AC links. The primary function of renewable sources is to generate electric energy, the excess of which can be accumulated

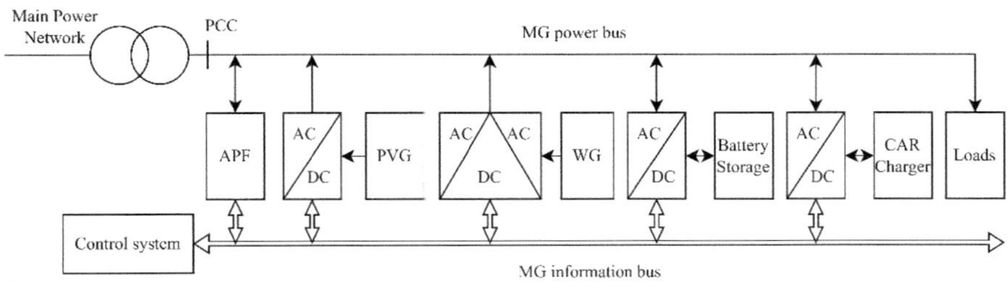

Figure 4.1 Distributed generation system and consumption with centralized or distributed control.

in battery storage for the own needs of consumers of the microgrid or transferred to the main power network. The shunt active filter (SAF) compensates for the inactive components of the microgrid's load current, depriving them of the main power network. However, due to the rather high cost of a separately installed SAF, distributed active filtering has become widespread, placing the compensation functions of higher harmonics and unwanted power components on the semiconductor converters of the microgrid with apparent power reserves [10, 11]. In this regard, it is preferable to represent the unbalance power as an orthogonal component with separate contributions to the apparent power for easy control of the unbalance power.

Electric vehicles (EVs) are a significant trend in the transport sector, which means a significant impact on the sustainability and efficiency of the grid. The electrical network with charging stations should be optimized in terms of interaction with other electrical equipment, and should use a standard microgrid interface to provide power quality [12].

Joint work between electric vehicles and renewable energy sources forces the adoption of new conditions for controlling electric energy, also aimed at the emergence of new opportunities, considering the limitations of the power grid [13]. The impact of electric vehicles in various modes aimed at stabilizing the voltage in the power grid and introducing renewable energy sources is considered in [14]. Innovative strategies for charging electric vehicles, supported by algorithms to maximize cooperation with renewable energy sources, are proposed in [15]. A particular control strategy of electric energy, which is supported by integrating electric vehicles and renewable energy sources with different power profiles, is proposed in [16]. The above studies do not mention how semiconductor converters must interact to perform the functions of distributed active filtering.

4.2 Improvement of the Multiphase Instantaneous Power Theory, Considering the Ratio of Transmission Line Resistances

Multiphase power supply system (Figure 4.2) is fully described by n dimensional vector of phase voltages $\mathbf{u}(t) = \|u_1(t) \quad u_2(t) \quad \ldots \quad u_n(t)\|^\wedge$ that are measured relative to the neutral wire and vector of line currents $\mathbf{i}(t) = \|i_1(t) \quad i_2(t) \quad \ldots \quad i_n(t)\|^\wedge$, where $^\wedge$ is the transpose sign.

According to the authors in [17], instant active current $\mathbf{i}_a(t)$ is part of the load current vector, which provides the exact value of instant load power as determined by the expression

$$p(t) = u_1(t)i_1(t) + u_2(t)i_2(t) + \cdots + u_n(t)i_n(t) = \mathbf{u}^\wedge(t)\mathbf{i}(t), \tag{4.1}$$

and causes minimal instantaneous loss of power in the transmission line. From the instantaneous active current definition we have

$$\mathbf{u}^\wedge(t)\mathbf{i}_a(t) = p(t). \tag{4.2}$$

The following expression determines the instantaneous power of losses

$$\Delta p(t) = i_1^2(t)r + i_2^2(t)r + \cdots + i_n^2 r + i_N^2(t)r_N,$$

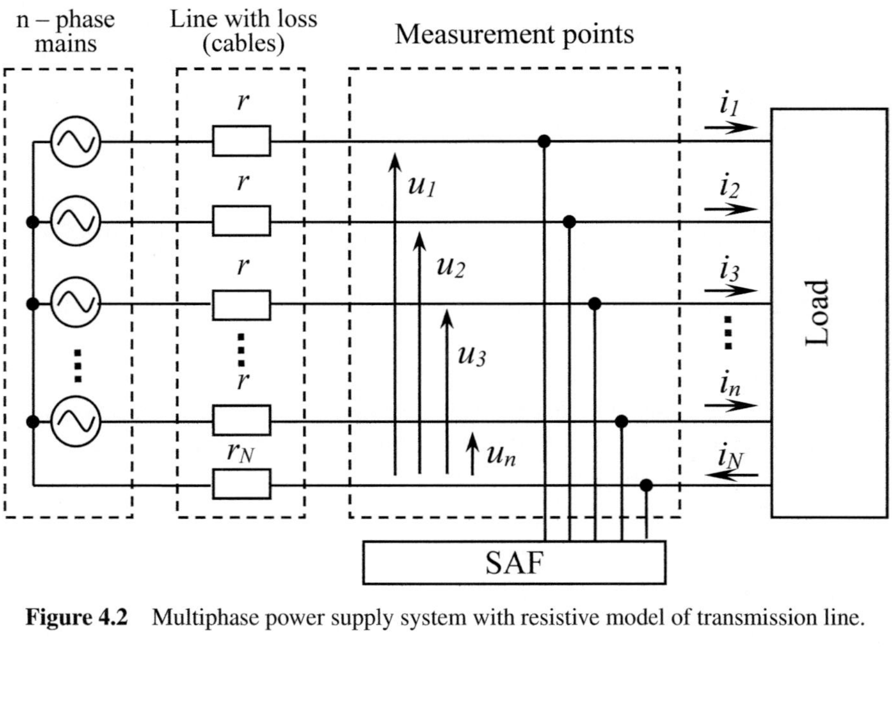

Figure 4.2 Multiphase power supply system with resistive model of transmission line.

where the current of the neutral can be represented in the form of

$$i_N(t) = i_1(t)+i_2(t)+\cdots+i_n(t) = \mathbf{i}^\wedge(t)\mathbf{j} = \mathbf{j}^\wedge\mathbf{i}(t);\mathbf{j}^\wedge = \| \; 1 \quad 1 \quad \ldots \quad 1 \; \|.$$

With this in mind, we will convert the expression for instantaneous loss power to matrix-vector form:

$$\Delta p(t) = \mathbf{i}^\wedge(t)\mathbf{i}(t)r+\mathbf{i}^\wedge(t)\mathbf{j}\mathbf{j}^\wedge\mathbf{i}(t)r_N = \mathbf{i}^\wedge(t)\mathbf{R}\mathbf{i}(t), \tag{4.3}$$

where $\mathbf{R} = r\mathbf{I}+r_N\mathbf{j}\mathbf{j}^\wedge = \mathbf{R}^\wedge$ is the matrix of resistances of losses of the transmission line is symmetrical concerning the main diagonal; **I** is the unit matrix of dimension $n\times n$.

We formulate an optimization problem to determine the instant active current [18]: for given quantities $\mathbf{R}$, $\mathbf{u}(t)$,$p(t)$ to determine the time vector-function $\mathbf{i}_a(t)$ that minimizes the instantaneous power loss ($\mathbf{i}_a^\wedge(t)\mathbf{R}\mathbf{i}_a(t) \to$ min) where subject to limitation is equality $\mathbf{u}^\wedge(t)\mathbf{i}_a(t) = p(t)$.

Consider solving a partial optimization problem for the case when $r_N = 0; \mathbf{R} = r\mathbf{I}$, that is, the instantaneous power loss is proportional to the square of the instantaneous norm of the line current vector, which we denote for this case $\mathbf{i}_1(t)$. Figure 4.3 illustrated the geometric content of this problem solution as actual minimization of the module or the length of the vector $\mathbf{i}_1(t)$. Since the left side of the given limit-equality (eqn (4.2)) relative to instantaneous power is a scalar product of vectors $\mathbf{u}(t)$ and $\mathbf{i}_1(t)$, it is satisfied with all vectors of currents ending in a plane perpendicular to the segment $p(t)$, deposited along the vector $\mathbf{u}(t)$.

The minimum length has a vector $\mathbf{i}_{1a}(t)$ perpendicular to the specified plane, that is, a parallel vector $\mathbf{u}(t)$. In the condition of parallelism

$$\mathbf{i}_{1a}(t) = g_1(t)\mathbf{u}(t), \tag{4.4}$$

we find the value of the proportionality coefficient $g_1(t) = p(t)/[\mathbf{u}^\wedge(t)\mathbf{u}(t)]$ from equality given in eqn (4.2). Substitution of this coefficient value in eqn (4.4) gives the formula of the instantaneous active current in the decomposition [19], which minimizes its instantaneous norm

$$\mathbf{i}_{1a}(t) = \frac{p(t)}{\mathbf{u}^\wedge(t)\mathbf{u}(t)}\mathbf{u}(t). \tag{4.5}$$

To minimize the instantaneous power loss (eqn (4.3)), we present it with a square of the vector norm $\bar{\mathbf{i}}(t) = \mathbf{R}^{1/2}\mathbf{i}(t)$:

$$\Delta p(t) = \bar{\mathbf{i}}^\wedge(t)\bar{\mathbf{i}}(t) = [\mathbf{R}^{1/2}\mathbf{i}(t)]^\wedge\mathbf{R}^{1/2}\mathbf{i}(t) =$$

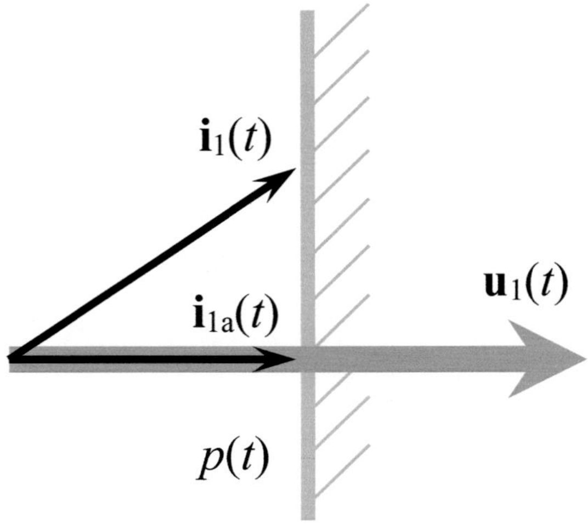

Figure 4.3 Geometric content of the solution of a partial optimization problem.

$$= \mathbf{i}^{\wedge}(t)(\mathbf{R}^{1/2})^{\wedge}\mathbf{R}^{1/2}\mathbf{i}(t) = \mathbf{i}^{\wedge}(t)\mathbf{R}\mathbf{i}(t). \tag{4.6}$$

Instant active power can be represented by the scalar product of vectors $\bar{\mathbf{i}}(t)$ and $\bar{\mathbf{u}}(t) = \mathbf{R}^{-1/2}\mathbf{u}(t)$:

$$p(t) = \bar{\mathbf{i}}^{\wedge}\bar{\mathbf{u}} = [\mathbf{R}^{1/2}\mathbf{i}(t)]^{\wedge}\mathbf{R}^{-1/2}\mathbf{u}(t) = \mathbf{i}^{\wedge}(t)(\mathbf{R}^{1/2})^{\wedge}\mathbf{R}^{-1/2}\mathbf{u}(t) =$$
$$= \mathbf{i}^{\wedge}(t)\,\mathbf{u}(t) = \mathbf{u}^{\wedge}(t)\,\mathbf{i}(t). \tag{4.7}$$

Similarly, the solution of a partial optimization problem, in general, minimization of the instantaneous power loss (eqn (4.6)) will take place subject to the proportionality of the introduced vectors:

$$\mathbf{R}^{1/2}\mathbf{i}_a(t) = k(t)\mathbf{R}^{-1/2}\mathbf{u}(t), \tag{4.8}$$

that, considering the value of the proportionality coefficient $k(t)$ found from eqn (4.2), contains an expression for the instantaneous active current of the multiphase power system [18]

$$\mathbf{i}_a(t) = \frac{p(t)}{\mathbf{u}^{\wedge}(t)\mathbf{R}^{-1}\mathbf{u}(t)}\mathbf{R}^{-1}\mathbf{u}(t). \tag{4.9}$$

Unlike the expression given above in eqn (4.5), the instantaneous active current (eqn (4.9)) minimizes the instantaneous power loss, not the norm of the line current vector. These expressions become identical with zero neutral resistance ($r_N = 0$) or zero-sum of instantaneous coordinate values of the phase voltage vector ($\mathbf{j}^{\wedge}\mathbf{u}(t) = 0$).

We apply the Cauchy–Schwartz inequality [25] for vectors $\bar{\mathbf{i}}(t)$ and $\bar{\mathbf{u}}(t)$, its scalar product is the instantaneous active power, and the square of the modified current vector norm is equal to the instantaneous power loss:

$$p^2(t) = [\bar{\mathbf{i}}^{\wedge}(t)\bar{\mathbf{u}}(t)]^2 \leq [\bar{\mathbf{i}}^{\wedge}(t)\bar{\mathbf{i}}(t)][\bar{\mathbf{u}}^{\wedge}(t)\bar{\mathbf{u}}(t)] = \Delta p(t)[\mathbf{u}^{\wedge}(t)\mathbf{R}^{-1}\mathbf{u}(t)]. \tag{4.10}$$

According to the resulting inequality, instant active power cannot exceed the square root on the right side; this is its maximum value, which, according to [17], is taken as instantaneous apparent power:

$$s(t) = p_{\mathrm{MAX}}(t) = \sqrt{[\mathbf{i}^{\wedge}(t)\mathbf{R}\mathbf{i}(t)][\mathbf{u}^{\wedge}(t)\mathbf{R}^{-1}\mathbf{u}(t)]}. \tag{4.11}$$

Find the physical content of the scalar value $\mathbf{u}^{\wedge}(t)\mathbf{R}^{-1}\mathbf{u}(t)$, which also appears in the definition of an instantaneous active current (eqn (4.9)). To do this, we substitute in this expression the vector of phase voltages, expressed through the vector of phase EMF $\mathbf{e}(t)$ and the vector of line currents according to the formula $\mathbf{u}(t) = \mathbf{e}(t) - \mathbf{R}\mathbf{i}(t)$:

$$\begin{aligned}
&\mathbf{u}(t) = \mathbf{e}(t) - \mathbf{R}\mathbf{i}(t) :\\
&\quad \mathbf{u}^{\wedge}(t)\mathbf{R}^{-1}\mathbf{u}(t) = \left[\mathbf{e}^{\wedge}(t) - \mathbf{i}^{\wedge}(t)\mathbf{R}\right]\mathbf{R}^{-1}[\mathbf{e}(t) - \mathbf{R}\mathbf{i}(t)] =\\
&\quad = \mathbf{e}^{\wedge}(t)\mathbf{R}^{-1}\mathbf{e}(t) - \mathbf{i}^{\wedge}(t)\mathbf{e}(t) - \mathbf{e}^{\wedge}(t)\mathbf{i}(t) + \mathbf{i}^{\wedge}(t)\mathbf{R}\mathbf{i}(t) =\\
&\quad = \mathbf{e}^{\wedge}(t)\mathbf{R}^{-1}\mathbf{e}(t) - 2[p(t) + \Delta p(t)] + \Delta p(t) = p_0(t) - 2p(t) - \Delta p(t)
\end{aligned}$$

where $p_0(t) = \mathbf{e}^{\wedge}(t)\mathbf{R}^{-1}\mathbf{e}(t) = \mathbf{e}^{\wedge}(t)\mathbf{i}_0(t)$ is the instantaneous short circuit power of the n-phase source on the transmission line with the specified resistive model; $\mathbf{i}_0(t) = \mathbf{R}^{-1}\mathbf{e}(t)$ is short circuit current of the n-phase source, which follows from the equality $\mathbf{e}(t) - \mathbf{R}\mathbf{i}_0(t) = \mathbf{0}$. In a properly functioning multiphase power system $p_0(t) >> p(t) >> \Delta p(t)$, therefore, it can be assumed that the expression under study is an instantaneous short circuit power of the n-phase source on the transmission line resistances:

$$\mathbf{u}^{\wedge}(t)\mathbf{R}^{-1}\mathbf{u}(t) \approx \mathbf{e}^{\wedge}(t)\mathbf{R}^{-1}\mathbf{e}(t) = p_0(t).$$

Considering the physical content of the second multiplier (eqn (4.11)), the formula for determining instant apparent power looks like this:

$$s(t) = \sqrt{\Delta p(t) p_0(t)}, \tag{4.12}$$

that is, the apparent instantaneous power of the multiphase power system is the average geometric mean of instantaneous current power loss $\Delta p(t)$ and instantaneous voltage short circuit power $p_0(t)$ [20, 21].

From the inequality (eqn (4.10)), it also follows that the instantaneous power loss $\Delta p(t)$ in the transmission line caused by the arbitrary current vector $\mathbf{i}\,(t)$ does not exceed the minimum possible value $\Delta p_{\mathrm{MIN}}(t)$ determined by the expression:

$$\Delta p(t) \geq \Delta p_{\mathrm{MIN}}(t) = p^2(t)/[\mathbf{u}^\wedge(t)\mathbf{R}^{-1}\mathbf{u}(t)] = p^2(t)/p_0(t). \qquad (4.13)$$

Find instantaneous power loss from active current (eqn (4.9)):

$$\begin{aligned}\Delta p_a(t) = \mathbf{i}_a^\wedge(t)\mathbf{R}\mathbf{i}_a(t) &= \left[\frac{p(t)}{\mathbf{u}^\wedge(t)\mathbf{R}^{-1}\mathbf{u}(t)}\right]^2 \left[\mathbf{R}^{-1}\mathbf{u}(t)\right]^\wedge \mathbf{R}\left[\mathbf{R}^{-1}\mathbf{u}(t)\right] = \\ &= \frac{p^2(t)}{\mathbf{u}^\wedge(t)\mathbf{R}^{-1}\mathbf{u}(t)} = \frac{p^2(t)}{p_0(t)} = \Delta p_{\mathrm{MIN}}(t) \qquad (4.14)\end{aligned}$$

So, according to the formula (eqn (4.13)), the instantaneous active current (eqn (4.9)) provides the minimum possible instantaneous power loss in the transmission line. Inactive instantaneous current, which must be compensated by active filtering is determined by the expression

$$\mathbf{i}_{na}(t) = \mathbf{i}(t) - \mathbf{i}_a(t). \qquad (4.15)$$

Instantaneous active power of inactive current (eqn (4.15)) is zero because

$$\mathbf{u}^\wedge(t)\mathbf{i}_{na}(t) = \mathbf{u}^\wedge(t)\mathbf{i}(t) - \mathbf{u}^\wedge(t)\mathbf{i}_a(t) = p(t) - p(t) = 0.$$

This makes it possible to build a system of shunt active filtration in real-time without electrical energy storage element and with the minimum possible instantaneous power loss in the transmission line. For its implementation, an active shunt filter (SAF) is installed directly on the load terminals (Figure 4.2) and supplies an inactive current (eqn (4.15)) to the load, depriving it of a multiphase source and the transmission line.

Decomposition (eqn (4.15)) of the instant load current vector causes the following components of instantaneous power losses:

$$\begin{aligned}\Delta p(t) &= \mathbf{i}^\wedge(t)\mathbf{R}\mathbf{i}(t) = \\ &= \mathbf{i}_a^\wedge(t)\mathbf{R}\mathbf{i}_a(t) + \mathbf{i}_{na}^\wedge(t)\mathbf{R}\mathbf{i}_a(t) + \mathbf{i}_a^\wedge(t)\mathbf{R}\mathbf{i}_{na}(t) + \mathbf{i}_{na}^\wedge(t)\mathbf{R}\mathbf{i}_{na}(t).\end{aligned}$$

The specific determination by (eqn (4.9)) of the instantaneous active current vector generates a unique property of equality of zero mutual instantaneous losses caused by current vectors $\mathbf{i}_a(t)$ and $\mathbf{i}_{na}(t)$ [18]. In fact, this is confirmed as follows:

$$\mathbf{i}_{na}^{\wedge}(t)\mathbf{R}\mathbf{i}_a(t) = \mathbf{i}_a^{\wedge}(t)\mathbf{R}\mathbf{i}_{na}(t) = \mathbf{i}_a^{\wedge}(t)\mathbf{R}\mathbf{i}(t) - \mathbf{i}_a^{\wedge}(t)\mathbf{R}\mathbf{i}_a(t) =$$
$$\frac{p(t)}{p_0(t)}\mathbf{u}^{\wedge}(t)\mathbf{R}^{-1}\mathbf{R}\mathbf{i}(t) - \left[\frac{p(t)}{p_0(t)}\right]^2 \mathbf{u}^{\wedge}(t)\mathbf{R}^{-1}\mathbf{R}\mathbf{R}^{-1}$$
$$\mathbf{u}(t) = \frac{p(t)}{p_0(t)}p(t) - \frac{p^2(t)}{p_0^2(t)}p_0(t) = 0.$$

So, even though, in general, the vectors $\mathbf{i}_a(t)$ and $\mathbf{i}_{na}(t)$ are not orthogonal since their scalar product is not zero, they generate zero mutual instantaneous power losses, because of which there is such a decomposition of the instantaneous power losses in the transmission line

$$\Delta p(t) = \mathbf{i}_a^{\wedge}(t)\mathbf{R}\mathbf{i}_a(t) + \mathbf{i}_{nc}^{\wedge}(t)\mathbf{R}\mathbf{i}_{na}(t) = \Delta p_a(t) + \Delta p_{na}(t). \tag{4.16}$$

The minimum instantaneous power loss in the transmission line, subject to instant load power $p(t)$ is $\Delta p_{\mathrm{MIN}}(t) = \mathbf{i}_a^{\wedge}(t)\mathbf{R}\mathbf{i}_a(t)$. Any difference between the source current vector and the instantaneous active current (eqn (4.9)) is accompanied by additional instantaneous losses

$$\Delta p_{na}(t) = \mathbf{i}_{na}^{\wedge}(t)\mathbf{R}\mathbf{i}_{na}(t) = \mathbf{i}_{na}^{\wedge}(t)\mathbf{i}_{na}(t)r + [\mathbf{j}^{\wedge}\mathbf{i}_{na}(t)]^2 r_N. \tag{4.17}$$

The efficiency of energy transmission process by the current $\mathbf{i}(t)$ at any time characterizes the instantaneous power factor [18]

$$\lambda(t) = p(t)/s(t) = p(t)/\sqrt{p_0(t)\Delta p(t)}, \tag{4.18}$$

which acquires a maximum value equal to unity, only for instant active current (eqn (4.9)). Indeed, in the substitution of the instantaneous power loss $\Delta p(t) = \Delta p_{\mathrm{MIN}}(t) = p^2(t)/p_0(t)$ caused by it in (eqn (9.18)), we obtain that $\lambda(t) = 1$.

This value of $\Delta p_{\mathrm{MIN}}(t)$ and instant apparent power definition (eqn (4.12)) are also satisfying the gain coefficient for the instant power loss, first introduced in [22]:

$$w(t) = \frac{\Delta p(t)}{\Delta p_{\mathrm{MIN}}(t)} = \frac{\Delta p(t)}{p^2(t)/p_0(t)} = \frac{s^2(t)}{p^2(t)} = \frac{1}{\lambda^2(t)}. \tag{4.19}$$

Hence the ratio for the instantaneous power factor

$$\lambda(t) = \frac{p(t)}{s(t)} = \frac{1}{\sqrt{w(t)}} = \sqrt{\frac{\Delta p_{\mathrm{MIN}}(t)}{\Delta p(t)}} = \sqrt{\frac{\mathbf{i}_a^{\wedge}(t)\mathbf{R}\mathbf{i}_a(t)}{\mathbf{i}^{\wedge}(t)\mathbf{R}\mathbf{i}(t)}}, \tag{4.20}$$

can be used to determine it experimentally through the instant power loss.

We will compare the instantaneous power losses for various control SAF strategies for forming multiphase source currents. The most common of them is using the reference vector of phase voltages with a different coefficient of the zero-sequence component weakening.

Let's find the optimal meaning of the active current (eqn (4.9)) reference vector. To do this, we introduce a normalized matrix of loss resistances $\overline{\mathbf{R}} = \mathbf{R}/r = \mathbf{I}{+}(r_N/r)\mathbf{j}\mathbf{j}^{\wedge}$ and, considering the value of the zero-sequence unit vector in the *n*-phase system $\mathbf{e}_0 = n^{-1/2}\mathbf{j}$, we will convert the expression

$$\overline{\mathbf{R}}^{-1}\mathbf{u}(t) = \left(\mathbf{I}{+}\frac{r_N}{r}\mathbf{j}\mathbf{j}^{\wedge}\right)^{-1}\mathbf{u}(t) = \left[\mathbf{I}{-}\frac{r_N}{r(1{+}r_N\mathbf{j}^{\wedge}\mathbf{j}/r)}\mathbf{j}\mathbf{j}^{\wedge}\right]\mathbf{u}(t) =$$

$$= \mathbf{u}(t){-}\frac{nr_N}{r{+}nr_N}\times\frac{\mathbf{j}^{\wedge}\mathbf{u}(t)}{n}\mathbf{j} = \mathbf{u}(t){-}\sigma[\mathbf{e}_0^{\wedge}\mathbf{u}(t)]\mathbf{e}_0 = \mathbf{u}(t){-}\sigma\mathbf{u}_0(t) = \mathbf{u}_\sigma(t),$$

where $\mathbf{u}_0(t) = [\mathbf{e}_0^{\wedge}\mathbf{u}(t)]\mathbf{e}_0 = [\mathbf{j}^{\wedge}\mathbf{u}(t)]\mathbf{j}/n$ is the reference voltage zero-sequence component; $\sigma = nr_N/(r{+}nr_N)$ is coefficient of its optimal weakening.

Considering the introduced designations, the formula of instant active current takes the form of:

$$\mathbf{i}_a(t) = \frac{p(t)}{\mathbf{u}^{\wedge}(t)\overline{\mathbf{R}}^{-1}\mathbf{u}(t)}\overline{\mathbf{R}}^{-1}\mathbf{u}(t) = \frac{p(t)}{\mathbf{u}^{\wedge}(t)\mathbf{u}_\sigma(t)}\mathbf{u}_\sigma(t). \tag{4.21}$$

By introducing an orthogonal to $\mathbf{u}_0(t)$ the component of the phase voltage vector $\mathbf{u}_\perp(t) = \mathbf{u}(t){-}\mathbf{u}_0(t)$, an instant active current can be represented as follows:

$$\mathbf{i}_a(t) = \frac{p(t)}{\mathbf{u}^{\wedge}(t)\mathbf{u}_\sigma(t)}\mathbf{u}_\sigma(t) = \frac{p(t)}{u_\perp^2(t) + (1{-}\sigma)u_0^2(t)}[\mathbf{u}_\perp(t) + (1{-}\sigma)\mathbf{u}_0(t)], \tag{4.22}$$

where $u_\perp^2(t) = \mathbf{u}_\perp^{\wedge}(t)\mathbf{u}_\perp(t); u_0^2(t) = \mathbf{u}_0^{\wedge}(t)\mathbf{u}_0(t)$ the squares of the corresponding vector norms. The current eqn (4.22)) should be formed in the transmission line of the multiphase system using the SAF to minimize the instantaneous power loss. In the multiphase theory of shunt active filtering such strategies for improving the instantaneous power factor are widely used. The strategy for minimizing the quadratic rate of current consumption [22] shapes the power supply current vector as follows:

$$\mathbf{i}_1(t) = \frac{p(t)}{\mathbf{u}^\wedge(t)\mathbf{u}(t)}\mathbf{u}(t) = \frac{p(t)}{u_\perp^2(t)+u_0^2(t)}[\mathbf{u}_\perp(t)+\mathbf{u}_0(t)]. \tag{4.23}$$

The neutral zero-current forming strategy [23] expresses the current consumed as follows:

$$\mathbf{i}_0(t) = \frac{p(t)}{\mathbf{u}_\perp^\wedge(t)\mathbf{u}_\perp(t)}\mathbf{u}_\perp(t) = \frac{p(t)}{u_\perp^2(t)}\mathbf{u}_\perp(t). \tag{4.24}$$

All three strategies (eqn (4.22)–(4.24)) can be covered by a generalized formula

$$\mathbf{i}_\gamma(t) = \frac{p(t)}{u_\perp^2(t)+\gamma u_0^2(t)}[\mathbf{u}_\perp(t)+\gamma\mathbf{u}_0(t)], \tag{4.25}$$

where the γ parameter takes on a value of $1-\sigma$ in strategy represented by eqn (4.22), 1 in strategy represented by eqn (4.23), and 0 in strategy represented by eqn (4.24).

Instant power loss from current generalized strategy

$$\begin{aligned}\Delta p_\gamma(t) &= \frac{p^2(t)}{[u_\perp^2(t)+\gamma u_0^2(t)]^2}[\mathbf{u}_\perp(t)+\gamma\mathbf{u}_0(t)]^\wedge(r\mathbf{I}+r_N\mathbf{j}\mathbf{j}^\wedge)[\mathbf{u}_\perp(t)+\gamma\mathbf{u}_0(t)] = \\ &= \frac{p^2(t)}{[u_\perp^2(t)+\gamma u_0^2(t)]^2}\{ru_\perp^2(t)+r\gamma^2u_0^2(t)+r_N\gamma^2[\mathbf{j}^\wedge\mathbf{u}(t)]^2\} = \\ &= \frac{rp^2(t)}{[u_\perp^2(t)+\gamma u_0^2(t)]^2}[u_\perp^2(t)+\gamma^2u_0^2(t)(r+nr_N)/r].\end{aligned}$$

Because $1-\sigma = 1-nr_N/(r+nr_N) = r/(r+nr_N)$ the functional dependence of the generalized instantaneous power loss from the γ parameter is as follows:

$$\frac{\Delta p_\gamma(t)}{rp^2(t)} = \frac{u_\perp^2(t)+\gamma^2(1-\sigma)^{-1}u_0^2(t)}{[u_\perp^2(t)+\gamma u_0^2(t)]^2}. \tag{4.26}$$

The minimization study of this functional dependence gives the optimal value $\gamma_0 = 1-\sigma$ which corresponds to the SAF strategy of formation the instant active current represented by eqn (4.21) with considered optimal weakening of the zero-sequence component.

The value of the minimum instantaneous power of losses will be taken in case of substituting the optimal value $\gamma_0 = 1-\sigma$ in eqn (4.26):

$$\Delta p_\gamma(t)\,|_{\,\gamma_0=1-\sigma} = \frac{rp^2(t)}{u_\perp^2(t) + (1-\sigma)u_0^2(t)} = \frac{p^2(t)}{\mathbf{u}^\wedge(t)\mathbf{R}^{-1}\mathbf{u}(t)} = \Delta p_{\mathrm{MIN}}(t),$$

which, as expected, is entirely the same as eqn (4.14). Thus, studies on the extremum of the analytical dependence of the instantaneous power losses from the γ parameter fully confirmed the validity of eqn (4.9) to determine the vector of instantaneous active current. It, in the presence of an instantaneous voltage zero-sequence component, is superior in terms of instantaneous power losses to strategies based on the orthogonal decomposition of load currents using *pq*-theory of instantaneous power and its modification, cross-vector theory, instantaneous *pqr* power theory, proportional-vector theory with deprivation of the voltage zero-sequence component.

Multiplying both parts of equality represented by eqn (4.16) by instant short circuit power, we get the decomposition of the square of instantaneous apparent power

$$s^2(t) = \Delta p(t)p_0(t) = p^2(t)+q^2(t), \tag{4.27}$$

where $q^2(t) = p_0(t)\Delta p_{na}(t) = [\mathbf{u}^\wedge(t)\mathbf{R}^{-1}\mathbf{u}(t)][\mathbf{i}_{na}^\wedge(t)\mathbf{R}\mathbf{i}_{na}(t)]$ is the square of instantaneous inactive power.

Taking into account eqn (4.22), we present the vector of inactive current in the matrix-vector form [18, 24]

$$\begin{aligned}\mathbf{i}_{na}(t) &= \mathbf{i}(t) - \mathbf{i}_a(t) = \mathbf{i}(t) - \frac{\mathbf{i}^\wedge(t)\mathbf{u}(t)}{\mathbf{u}^\wedge(t)\mathbf{u}_\sigma(t)}\mathbf{u}_\sigma(t) = \\ &= \frac{1}{\mathbf{u}^\wedge(t)\mathbf{u}_\sigma(t)}\left\{\mathbf{i}(t)\left[\mathbf{u}_\sigma^\wedge(t)\mathbf{u}(t)\right] - \mathbf{u}_\sigma(t)\left[\mathbf{i}^\wedge(t)\mathbf{u}(t)\right]\right\} = \\ &= \frac{1}{\mathbf{u}^\wedge(t)\mathbf{u}_\sigma(t)}\left\{\left[\mathbf{i}(t)\mathbf{u}_\sigma^\wedge(t) - \mathbf{u}_\sigma(t)\mathbf{i}^\wedge(t)\right]\mathbf{u}(t)\right\} = \frac{1}{\mathbf{u}^\wedge(t)\mathbf{u}_\sigma(t)}\mathbf{Q}(t)\mathbf{u}(t),\end{aligned} \tag{4.28}$$

where where $\mathbf{Q}(t) = \mathbf{i}(t)\mathbf{u}_\sigma^\wedge(t) - \mathbf{u}_\sigma(t)\mathbf{i}^\wedge(t)$ is a skew matrix with a zero main diagonal is different from zero only if the vectors $\mathbf{i}(t), \mathbf{u}_\sigma(t)$ are disproportionate.

So, we have the following decomposition formula of the line current vector into an instant active current (eqn (4.21)) and an instantaneous inactive current according to the formula represented by eqn (4.28):

$$\mathbf{i}(t) = \frac{p(t)}{\mathbf{u}^\wedge(t)\mathbf{u}_\sigma(t)}\mathbf{u}_\sigma(t)+\frac{1}{\mathbf{u}^\wedge(t)\mathbf{u}_\sigma(t)}\mathbf{Q}(t)\mathbf{u}(t). \tag{4.29}$$

It is easy to establish that the quadratic form of the phase voltage vector with the matrix $\mathbf{Q}(t)$ is zero at any given time:

$$\mathbf{u}^{\wedge}(t)\mathbf{Q}(t)\mathbf{u}(t) = \mathbf{u}^{\wedge}(t)[\mathbf{i}(t)\mathbf{u}_{\sigma}^{\wedge}(t) - \mathbf{u}_{\sigma}(t)\mathbf{i}^{\wedge}(t)]\mathbf{u}(t) =$$
$$= [\mathbf{u}^{\wedge}(t)\mathbf{i}(t)] \times [\mathbf{u}_{\sigma}^{\wedge}(t)\mathbf{u}(t)] - [\mathbf{u}^{\wedge}(t)\mathbf{u}_{\sigma}(t)] \times [\mathbf{i}^{\wedge}(t)\mathbf{u}(t)] = 0.$$

Considering this property and eqn (4.28), we will establish a connection between the square of instant inactive power and the matrix $\mathbf{Q}(t)$:

$$q^2(t) = [\mathbf{u}^{\wedge}(t)\mathbf{R}^{-1}\mathbf{u}(t)][\mathbf{i}_{na}^{\wedge}(t)\mathbf{R}\mathbf{i}_{na}(t)] =$$
$$= [\mathbf{u}^{\wedge}(t)\mathbf{R}^{-1}\mathbf{u}(t)][\mathbf{i}^{\wedge}(t)\mathbf{R}\mathbf{i}_{na}(t)] - [\mathbf{u}^{\wedge}(t)\mathbf{R}^{-1}\mathbf{u}(t)][\mathbf{i}_{a}^{\wedge}(t)\mathbf{R}\mathbf{i}_{na}(t)] = \quad (4.30)$$
$$= \frac{\mathbf{i}^{\wedge}(t)\mathbf{R}\mathbf{Q}(t)\mathbf{u}(t)}{r} - \frac{p(t)\left[\mathbf{u}^{\wedge}(t)\overline{\mathbf{R}}^{-1}\mathbf{R}\mathbf{Q}(t)\mathbf{u}(t)\right]}{[\mathbf{u}^{\wedge}(t)\mathbf{u}_{\sigma}(t)]\,r} = \mathbf{i}^{\wedge}(t)\overline{\mathbf{R}}\mathbf{Q}(t)\mathbf{u}(t).$$

We get the quadratic components $q^2(t)$ using Laplace equality [25] for vectors $\overline{\mathbf{R}}^{1/2}\mathbf{i}(t) = \overline{\mathbf{1}} = \|\bar{i}_1\bar{i}_2\ldots\bar{i}_n\|^{\wedge}$ and $\overline{\mathbf{R}}^{-1/2}\mathbf{u}(t) = \overline{\mathbf{u}} = \|\bar{u}_1\bar{u}_2\ldots \bar{u}_n\|^{\wedge}$:

$$\left(\sum_{j=1}^{n}\bar{i}_j^2\right)\times\left(\sum_{j=1}^{n}\bar{u}_j^2\right)-\left(\sum_{j=1}^{n}\bar{i}_j\bar{u}_j\right)^2 = (\bar{\mathbf{i}}^{\wedge}\bar{\mathbf{i}})\times(\overline{\mathbf{u}}^{\wedge}\overline{\mathbf{u}}) - (\bar{\mathbf{i}}^{\wedge}\overline{\mathbf{u}})^2 =$$
$$= \mathbf{i}^{\wedge}(t)\overline{\mathbf{R}}\mathbf{i}(t)\times\mathbf{u}^{\wedge}(t)\overline{\mathbf{R}}^{-1}\mathbf{u}(t) - [\mathbf{i}^{\wedge}(t)\overline{\mathbf{R}}^{1/2}\overline{\mathbf{R}}^{-1/2}\mathbf{u}(t)]^2 =$$
$$= s^2(t) - p^2(t) = q^2(t) = \frac{1}{2}\sum_{j=1}^{n}\sum_{k=1}^{n}(\bar{i}_j\bar{u}_k - \bar{i}_k\bar{u}_j)^2.$$

The last double sum is completely coincides with the sum of the squares of the elements of the matrix

$$\overline{\mathbf{Q}} = \overline{\mathbf{i}}\overline{\mathbf{u}}^{\wedge} - \overline{\mathbf{u}}\overline{\mathbf{i}}^{\wedge} =$$
$$= \begin{Vmatrix} \bar{i}_1\bar{u}_1 - \bar{i}_1\bar{u}_1 & \bar{i}_1\bar{u}_2 - \bar{i}_2\bar{u}_1 & \ldots & \bar{i}_1\bar{u}_n - \bar{i}_n\bar{u}_1 \\ \bar{i}_2\bar{u}_1 - \bar{i}_1\bar{u}_2 & \bar{i}_2\bar{u}_2 - \bar{i}_2\bar{u}_2 & \ldots & \bar{i}_2\bar{u}_n - \bar{i}_n\bar{u}_2 \\ \ldots & \ldots & \ldots & \ldots \\ \bar{i}_n\bar{u}_1 - \bar{i}_1\bar{u}_n & \bar{i}_n\bar{u}_2 - \bar{l}_2\bar{u}_n & \ldots & \bar{i}_n\bar{u}_n - \bar{i}_n\bar{u}_n \end{Vmatrix} =$$
$$= \begin{Vmatrix} 0 & \bar{q}_{12} & \ldots & \bar{q}_{1n} \\ -\bar{q}_{12} & 0 & \ldots & \bar{q}_{2n} \\ \ldots & \ldots & \ldots & \ldots \\ -\bar{q}_{1n} & -\bar{q}_{2n} & \ldots & 0 \end{Vmatrix}, \quad (4.31)$$

which is also skew matrix with a zero-main diagonal. Considering this, the square of instant inactive power consists of the sum of the squares of the elements of the matrix $\overline{\mathbf{Q}}$ located above the main diagonal:

$$q^2(t) = \sum_{j=1}^{n} \sum_{k>j}^{n} \overline{q}_{jk}^2. \tag{4.32}$$

The number of such elements, depending on the dimension of the multiphase system, is equal to $m = n(n-1)/2$. Consequently, the quadratic components of instant inactive power are *m* elements of the matrix $\overline{\mathbf{Q}} = \overline{\mathbf{R}}^{1/2}\mathbf{Q}(t)\overline{\mathbf{R}}^{-1/2}$ located above the main diagonal.

Using the decomposition of squares of instant powers (eqn (4.27)) and definitions given by eqn (4.30) and eqn (4.32), we will have the following series of calculated formulas for the gain coefficient by the power of instant losses:

$$\begin{aligned} \omega(t) &= 1 + \frac{q^2(t)}{p^2(t)} = 1 + \frac{\mathbf{i}(t)^{\wedge}\overline{\mathbf{R}}\mathbf{Q}(t)\mathbf{u}(t)}{p^2(t)} = \\ &= 1 + \frac{\overline{\mathbf{i}}^{\wedge}\overline{\mathbf{Q}}\overline{\mathbf{u}}}{p^2(t)} = 1 + \sum_{j=1}^{n} \sum_{k>j}^{n} \overline{q}_{jk}^2 / p^2(t). \end{aligned} \tag{4.33}$$

Consequently, each of the nonzero elements of the matrix $\overline{\mathbf{Q}} = \overline{\mathbf{R}}^{1/2}\mathbf{Q}(t)$ $\overline{\mathbf{R}}^{-1/2}$ corresponds to the component of the instantaneous power of losses in the transmission line.

As already noted, for three-dimensional vectors $\mathbf{u}(t) = \|u_A(t) \quad u_B(t) \quad u_C(t)\|^{\wedge}$ and currents $\mathbf{i}(t) = \|i_A(t) \quad i_B(t) \quad i_C(t)\|^{\wedge}$ that completely characterize the electromagnetic process in the cross-section $\langle a, B, c \rangle$ of the power supply system (Figure 4.4), a mathematical operation of the vector product is defined predetermines many features in determining the concepts of instantaneous power theory.

Instant load and loss power is determined according to the modified theory of instantaneous power of multiphase systems when $n = 3$ and $\mathbf{j}^{\wedge} = \| \; 1 \quad 1 \quad 1 \; \|$:

$$p(t) = \mathbf{u}^{\wedge}(t)\mathbf{i}(t) = u_a(t)i_a(t) + u_b(t)i_b(t) + u_c(t)i_c(t);$$
$$\Delta p(t) = \mathbf{i}^{\wedge}(t)\mathbf{R}\mathbf{i}(t) = \mathbf{i}^{\wedge}(t)\left(r\mathbf{I} + r_N\mathbf{j}\mathbf{j}^{\wedge}\right)\mathbf{i}(t) =$$

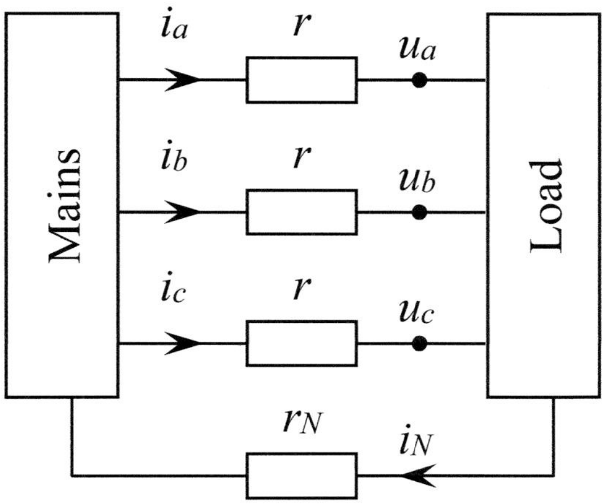

Figure 4.4 Three-phase four-wire power system using transmission line resistive model.

$$= \begin{Vmatrix} i_a(t) \\ i_b(t) \\ i_c(t) \end{Vmatrix}^{\wedge} \begin{Vmatrix} r+r_N & r_N & r_N \\ r_N & r+r_N & r_N \\ r_N & r_N & r+r_N \end{Vmatrix} \begin{Vmatrix} i_a(t) \\ i_b(t) \\ i_c(t) \end{Vmatrix}.$$

We will confirm the solution obtained in the previous subsection of the optimization problem of determining the instant active current $\mathbf{i}_a(t)$, which minimizes the instantaneous power of losses $\mathbf{i}_a^{\wedge}(t)\mathbf{R}\mathbf{i}_a(t) \to \min$, subject to limitation equality $p(t) - \mathbf{u}^{\wedge}\mathbf{i}_a(t) = 0$ by the Lagrange multiplier method [24]. To do this, form the Lagrange function

$$L(t) = \mathbf{i}_a^{\wedge}(t)\mathbf{R}\mathbf{i}_a(t) + \mu(t)[p(t) - \mathbf{u}^{\wedge}(t)\mathbf{i}_a(t)] =$$
$$= \mu(t)p(t) + [\mathbf{i}_a^{\wedge}(t)\mathbf{R} - \mu(t)\mathbf{u}^{\wedge}(t)]\mathbf{i}_a(t),$$

and differentiate it by all coordinates $i_{ak}(t)$ of the vector $\mathbf{i}_a(t)(k = a, b, c)$:

$$\frac{\partial L}{\partial i_{ak}} = [\mathbf{i}_a^{\wedge}(t)\mathbf{R} - \mu(t)\mathbf{u}^{\wedge}(t)]\mathbf{j}_k + \mathbf{j}_k^{\wedge}\mathbf{R}\mathbf{i}_a(t) =$$
$$= [\mathbf{i}_a^{\wedge}(t)\mathbf{R} - \mu(t)\mathbf{u}^{\wedge}(t)]\mathbf{j}_k + \mathbf{i}_a^{\wedge}(t)\mathbf{R}\mathbf{j}_k = [2\mathbf{i}_a^{\wedge}(t)\mathbf{R} - \mu(t)\mathbf{u}^{\wedge}(t)]\mathbf{j}_k,$$

where $\mathbf{j}_a = \begin{Vmatrix} 1 & 0 & 0 \end{Vmatrix}^{\wedge}$; $\mathbf{j}_b = \begin{Vmatrix} 0 & 1 & 0 \end{Vmatrix}^{\wedge}$; $\mathbf{j}_c = \begin{Vmatrix} 0 & 0 & 1 \end{Vmatrix}^{\wedge}$.

The system of equations for determining the instant active current takes the form of:

$$\begin{cases} [2\mathbf{i}_a^{\wedge}(t)\mathbf{R} - \mu(t)\mathbf{u}^{\wedge}(t)]\mathbf{j}_a = 0; \\ [2\mathbf{i}_a^{\wedge}(t)\mathbf{R} - \mu(t)\mathbf{u}^{\wedge}(t)]\mathbf{j}_b = 0; \\ [2\mathbf{i}_a^{\wedge}(t)\mathbf{R} - \mu(t)\mathbf{u}^{\wedge}(t)]\mathbf{j}_c = 0; \\ \mathbf{u}^{\wedge}(t)\mathbf{i}_a(t) = p(t). \end{cases} \quad (4.34)$$

From the first three equations of eqn (4.34), it follows that

$$\mathbf{i}_a(t) = \frac{\mu(t)}{2}\mathbf{R}^{-1}\mathbf{u}(t) = \mathrm{c}(t)\mathbf{R}^{-1}\mathbf{u}(t) = \mathrm{c}(t)\mathbf{i}_0(t),$$

where $\mathbf{i}_0(t) = \mathbf{R}^{-1}\mathbf{u}(t)$ is the vector of short-circuit currents of a three-phase source that satisfies the condition $\mathbf{u}(t) - \mathbf{R}\mathbf{i}_0(t) = \mathbf{0}$.

The scalar multiplier $\mathrm{c}(t)$ is determined from the fourth equation (eqn (4.34)) in the case of substitution $\mathbf{i}_a(t) = \mathrm{c}(t)\mathbf{R}^{-1}\mathbf{u}(t)$ in the form of

$$\mathrm{c}(t) = \frac{\mathbf{u}^\wedge(t)\mathbf{i}(t)}{\mathbf{u}^\wedge(t)\mathbf{R}^{-1}\mathbf{u}(t)} = \frac{p(t)}{\mathbf{u}^\wedge(t)\mathbf{i}_0(t)} = \frac{p(t)}{p_0(t)}, \tag{4.35}$$

where $p_0(t) = \mathbf{u}^\wedge(t)\mathbf{i}_0(t)$ is the instantaneous short-circuit power of the three-phase four-wire power supply system.

So, in another way, it is confirmed that in a three-phase four-wire power system, an instant active current may be defined as part of the short-circuit current vector of a three-phase source, which is equal to the ratio of instant load power to instant short-circuit power:

$$\mathbf{i}_a(t) = \frac{p(t)}{p_0(t)}\mathbf{i}_0(t). \tag{4.36}$$

Similarly eqn (4.21), when $n = 3$ instantaneous active current can be represented proportionally to the reference voltage vector:

$$\mathbf{i}_a(t) = \frac{p(t)}{\mathbf{u}^\wedge(t)\mathbf{u}_\sigma(t)}\mathbf{u}_\sigma(t), \tag{4.37}$$

where $\mathbf{u}_\sigma(t) = r\mathbf{R}^{-1}\mathbf{u}(t) = \overline{\mathbf{R}}^{-1}\mathbf{u}(t) = (\mathbf{I} + \frac{r_N}{r}\mathbf{j}\mathbf{j}^\wedge)^{-1}\mathbf{u}(t) = \mathbf{u}(t) - \sigma\mathbf{u}_0(t)$ is phase voltage vector with an optimally gain coefficient of the zero-sequence $\mathbf{u}_0(t) = u_0(t)\mathbf{j}$; $u_0(t) = [u_a(t) + u_b(t) + u_c(t)]/3$; $\sigma = 3r_N/(r + 3r_N)$ is the coefficient of optimal weakening of the zero-sequence component of the reference vector of phase voltages, which provides a minimum loss power in the four-wire transmission line [26, 27].

Instantaneous active current by eqn (4.37) causes neutral current

$$i_N(t) = \mathbf{j}^\wedge\mathbf{i}_a(t) = \frac{p(t)[\mathbf{j}^\wedge\mathbf{u}(t) - \sigma\mathbf{j}^\wedge\mathbf{j}u_0(t)]}{\mathbf{u}^\wedge(t)[\mathbf{u}(t) - \sigma\mathbf{u}_0(t)]} =$$
$$= \frac{3p(t)u_0(t)}{[1 + 3(r_N/r)][u_a^2(t) + u_b^2(t) + u_c^2(t)] - 9(r_N/r)u_0^2(t)}, \tag{4.38}$$

where the value is entirely the same as that obtained in [27].

We use specific mathematical operations on three-dimensional vectors to determine the direct formula of an inactive current of a three-phase four-wire power system. Double vector product formula [25]

$$\mathbf{a}\times(\mathbf{b}\times\mathbf{c}) = (\mathbf{a}^\wedge\mathbf{c})\mathbf{b}-(\mathbf{a}^\wedge\mathbf{b})\mathbf{c}$$

in the case, when $\mathbf{a} = \mathbf{u}(t);\mathbf{b} = \mathbf{i}(t);\mathbf{c} = \mathbf{R}^{-1}\mathbf{u}(t)$, takes the form of

$$\mathbf{u}(t) \times [\mathbf{i}(t)\times\mathbf{R}^{-1}\mathbf{u}(t)] = [\mathbf{u}^\wedge(t)\mathbf{R}^{-1}\mathbf{u}(t)]\mathbf{i}(t) - [\mathbf{u}^\wedge(t)\mathbf{i}(t)]\mathbf{R}^{-1}\mathbf{u}(t).$$

The last expression follows the formula for current decomposition into instant active and instant inactive currents [24]:

$$\mathbf{i}(t) = \frac{\mathbf{u}^\wedge(t)\mathbf{i}(t)}{\mathbf{u}^\wedge(t)\mathbf{R}^{-1}\mathbf{u}(t)}+\frac{\mathbf{u}(t) \times [\mathbf{i}(t)\times\mathbf{R}^{-1}\mathbf{u}(t)]}{\mathbf{u}^\wedge(t)\mathbf{R}^{-1}\mathbf{u}(t)} = \mathbf{i}_a(t)+\mathbf{i}_{na}(t), \quad (4.39)$$

where the instant inactive current is determined by a direct formula

$$\mathbf{i}_{na}(t) = \frac{\mathbf{u}(t) \times [\mathbf{i}(t)\times\mathbf{u}_\sigma(t)]}{\mathbf{u}^\wedge(t)\mathbf{u}_\sigma(t)} = \frac{\mathbf{u}(t)\times\mathbf{q}_\sigma(t)}{\mathbf{u}^\wedge(t)\mathbf{u}_\sigma(t)}. \quad (4.40)$$

A vector value $\mathbf{q}_\sigma(t) = \mathbf{i}(t)\times\mathbf{u}_\sigma(t) = \mathbf{i}(t)\times\overline{\mathbf{R}}^{-1}\mathbf{u}(t)$ is an instantaneous vector inactive power that meets the definition of active instant current (eqn (4.37)) in decomposition (eqn (4.39)). In contrast to the proposal in [28], the transmission line resistance factor is taken into account here. The order of vector factors in $\mathbf{q}_\sigma(t)$ corresponds to one proposed in [29] since it is this order provides the correct sign of reactive power at symmetrical sinusoidal currents and voltages.

Components of the determinant of the Gram matrix [25]

$$(\mathbf{a}^\wedge\mathbf{a})(\mathbf{b}^\wedge\mathbf{b}) = (\mathbf{a}^\wedge\mathbf{b})^2+(\mathbf{a}\times\mathbf{b})^\wedge(\mathbf{a}\times\mathbf{b}), \quad (4.41)$$

when $\mathbf{a} = \overline{\mathbf{R}}^{1/2}\mathbf{i}(t);\mathbf{b} = \overline{\mathbf{R}}^{-1/2}\mathbf{u}(t)$, acquire values:

$$\begin{aligned}
\mathbf{a}^\wedge\mathbf{a} &= [\overline{\mathbf{R}}^{1/2}\mathbf{i}(t)]^\wedge\overline{\mathbf{R}}^{1/2}\mathbf{i}(t) = \mathbf{i}(t)^\wedge\overline{\mathbf{R}}\mathbf{i}(t) = r\Delta p(t);\\
\mathbf{b}^\wedge\mathbf{b} &= [\overline{\mathbf{R}}^{-1/2}\mathbf{u}(t)]^\wedge\overline{\mathbf{R}}^{-1/2}\mathbf{u}(t) = \mathbf{u}(t)^\wedge\overline{\mathbf{R}}^{-1}\mathbf{u}(t) = p_0(t)/r;\\
\mathbf{a}^\wedge\mathbf{b} &= [\overline{\mathbf{R}}^{1/2}\mathbf{i}(t)]^\wedge\overline{\mathbf{R}}^{-1/2}\mathbf{u}(t) = \mathbf{u}(t)^\wedge\mathbf{i}(t) = p(t);\\
\mathbf{a}\times\mathbf{b} &= \overline{\mathbf{R}}^{1/2}\mathbf{i}(t)\times\overline{\mathbf{R}}^{-1/2}\mathbf{u}(t) = \overline{\mathbf{q}}(t) = \overline{\mathbf{R}}^{1/2}\mathbf{q}_\sigma(t)\overline{\mathbf{R}}^{-1/2} =\\
&= \|\overline{q}_a(t)\overline{q}_b(t)\overline{q}_c(t)\|^\wedge;\\
(\mathbf{a}\times\mathbf{b})^\wedge(\mathbf{a}\times\mathbf{b}) &= \overline{\mathbf{q}}^\wedge(t)\overline{\mathbf{q}}(t) = \overline{q}_a^2(t)+\overline{q}_b^2(t)+\overline{q}_c^2(t).
\end{aligned}$$

The left side of the eqn (4.41) is the square of instant apparent power determined by the formula represented by eqn (4.12):

$$s(t) = \sqrt{p_0(t)\Delta p(t)} = \sqrt{\mathbf{u}^\wedge(t)\mathbf{R}^{-1}\mathbf{u}(t)\times\mathbf{i}^\wedge(t)\mathbf{R}\mathbf{i}(t)}. \quad (4.42)$$

If zero-sequence components of current and voltage vectors or neutral wire resistance are equal to zero, eqn (4.42) passes into the formula in [28].

After substituting all values in the eqn (4.41), we have a decomposition of the square of apparent instantaneous power

$$s^2(t) = p^2(t)+q^2(t) = p^2(t)+\overline{\mathbf{q}}^\wedge(t)\overline{\mathbf{q}}(t) = p^2(t)+\overline{q}_a^2(t)+\overline{q}_b^2(t)+\overline{q}_c^2(t). \quad (4.43)$$

Thus, the inactive instantaneous power of a three-phase four-wire system has three quadratic components, the definition formulas of which coincide with eqn (4.31). Since $\overline{\mathbf{q}}^\wedge(t)\overline{\mathbf{q}}(t)\geq 0; p^2(t)\leq p_0(t)\Delta p(t)$ and the instantaneous power factor $\lambda(t) = p(t)/\sqrt{p_0(t)\Delta p(t)}$, which is determined based on instant apparent power according to the eqn (4.42), does not exceed a single value. Unlike the definition formula [30], it acquires a unit value only in the case of proportionality of vectors $\mathbf{i}(t), \mathbf{R}^{-1}\mathbf{u}(t)$, since at the same time $\mathbf{q}_\sigma(t) = \mathbf{i}(t)\times\mathbf{u}_\sigma(t) = [\mathbf{i}(t)\times\mathbf{R}^{-1}\mathbf{u}(t)]/r = \mathbf{0}$; $\overline{\mathbf{q}}(t) = \mathbf{0}$ and from the eqn (4.43) it follows that $p^2(t) = p_0(t)\Delta p(t)$.

We will introduce the instantaneous power loss gain coefficient as the ratio of instantaneous power loss from the flow of load current $\mathbf{i}(t)$ and the considered active current, providing the same instantaneous load power $p(t)$:

$$w(t) = \frac{\mathbf{i}^\wedge(t)\mathbf{R}\mathbf{i}(t)}{\mathbf{i}_a^\wedge(t)\mathbf{R}\mathbf{i}_a(t)}.$$

In general, the value is given as follows:

$$w(t) = \frac{\Delta p(t)}{\mathrm{c}^2(t)[\mathbf{R}^{-1}\mathbf{u}(t)]^\wedge\mathbf{R}\mathbf{i}_0(t)} = \frac{\Delta p(t)p_0(t)}{p^2(t)} = \frac{1}{\lambda^2(t)}. \quad (4.44)$$

Thus, the instantaneous power losses gain is equal to the minus second power of instantaneous power factor.

4.3 Development of Integral Power Theory of Multiphase Power Supply Systems

The integral indicators of multiphase power supply systems are obtained through the averaging of instantaneous indicators, as a rule, at the voltage period *T* of the multiphase network. The load active power is

$$P = \frac{1}{T}\int_0^T p(t)dt = \frac{1}{T}\int_0^T \mathbf{u}^\wedge(t)\mathbf{i}(t)dt = \mathbf{u}\circ\mathbf{i} = \frac{1}{T}\int_0^T \bar{\mathbf{u}}^\wedge(t)\bar{\mathbf{i}}(t)dt = \bar{\mathbf{u}}\circ\bar{\mathbf{i}}, \tag{4.45}$$

where the symbol $\circ$ indicates the integral scalar product of the corresponding periodic time vectors. The power losses is the square of the integral norm of the modified current vector

$$\Delta P = \frac{1}{T}\int_0^T \Delta p(t)dt = \frac{1}{T}\int_0^T \mathbf{i}^\wedge(t)\mathbf{R}\mathbf{i}(t)dt = \frac{1}{T}\int_0^T \bar{\mathbf{i}}^\wedge(t)\bar{\mathbf{i}}(t)dt = \bar{\mathbf{i}}\circ\bar{\mathbf{i}}. \tag{4.46}$$

For vectors $\bar{\mathbf{u}}(t) = \mathbf{R}^{-1/2}\mathbf{u}(t)$ and $\bar{\mathbf{i}}(t) = \mathbf{R}^{1/2}\mathbf{i}(t)$, that satisfying constraint $(\mathbf{R}^{-1/2}\mathbf{u})\circ(\mathbf{R}^{1/2}\mathbf{i}) = P$, let us write the Cauchy–Schwartz inequality [25]:

$$[(\mathbf{R}^{-1/2}\mathbf{u})\circ(\mathbf{R}^{1/2}\mathbf{i})]^2 \leq [(\mathbf{R}^{-1/2}\mathbf{u})\circ(\mathbf{R}^{-1/2}\mathbf{u})] \times [(\mathbf{R}^{1/2}\mathbf{i})\circ(\mathbf{R}^{1/2}\mathbf{i})].$$

This leads to inequality in the active power of the load

$$P = (\mathbf{R}^{-1/2}\mathbf{u})\circ(\mathbf{R}^{1/2}\mathbf{i}) \leq \sqrt{[(\mathbf{R}^{-1/2}\mathbf{u})\circ(\mathbf{R}^{-1/2}\mathbf{u})] \times [(\mathbf{R}^{1/2}\mathbf{i})\circ(\mathbf{R}^{1/2}\mathbf{i})]}, \tag{4.47}$$

and the formula for the apparent power of the multiphase power system [31–33] as the maximum active power on the right side of the inequality (eqn (4.47)):

$$S = P_{\max} = \sqrt{[(\mathbf{R}^{-1/2}\mathbf{u})\circ(\mathbf{R}^{-1/2}\mathbf{u})] \times [(\mathbf{R}^{1/2}\mathbf{i})\circ(\mathbf{R}^{1/2}\mathbf{i})]} =$$
$$= \sqrt{\frac{1}{T}\int_0^T \mathbf{u}^\wedge(t)\mathbf{R}^{-1}\mathbf{u}(t)dt \times \frac{1}{T}\int_0^T \mathbf{i}^\wedge(t)\mathbf{R}\mathbf{i}(t)dt}. \tag{4.48}$$

The resulting apparent power formula contains a multiplier as current power losses itself, not the average quadratic value of currents. It is valid for an arbitrary ratio between the active resistances of the transmission line wires. In the absence of zero-sequence components in phase voltage vectors and line currents, or at zero neutral resistance, the eqn (4.48) passes into the F. Buchholz formula of apparent power.

The first subjugate expression multiplier (eqn (4.48)) is the voltage short circuit power of a multiphase source averaged on period T

$$\frac{1}{T}\int_0^T \mathbf{u}^\wedge(t)\mathbf{R}^{-1}\mathbf{u}(t)dt = \frac{1}{T}\int_0^T p_0(t)dt = P_0, \tag{4.49}$$

considering that the apparent power of the multiphase system can be represented as an average geometric value of allowable power loss in transmission line and short-circuit power of multiphase source in resistances of the transmission line [33, 34]:

$$S = \sqrt{\Delta P P_0}. \tag{4.50}$$

The matrix **R** affects the optimal active current value in the presence of voltage zero-sequence component.

$$\mathbf{u}(t) = \mathbf{u}_\perp(t) + \mathbf{u}_0(t). \tag{4.51}$$

Considering the value of $\mathbf{R}^{-1}\mathbf{u}(t) = [\mathbf{u}_\perp(t) + (1-\sigma)\mathbf{u}_0(t)]/r$ the voltage integral from the expression given in eqn (4.49), it takes the form of

$$\begin{aligned}
\frac{1}{T}\int_0^T \mathbf{u}^\wedge(t)\mathbf{R}^{-1}\mathbf{u}(t)dt &= \frac{1}{Tr}\int_0^T [\mathbf{u}_\perp(t)+\mathbf{u}_0(t)]^\wedge[\mathbf{u}_\perp(t) + (1-\sigma)\mathbf{u}_0(t)]dt = \\
&= \frac{1}{Tr}\int_0^T \mathbf{u}_\perp^\wedge(t)\,\mathbf{u}_\perp(t)\,dt + (1-\sigma)\frac{1}{Tr}\int_0^T \mathbf{u}_0^\wedge(t)\mathbf{u}_0(t)]dt = \\
&= \frac{1}{r}[U_\perp^2 + (1-\sigma)U_0^2],
\end{aligned} \tag{4.52}$$

where $U_\perp^2 = \frac{1}{T}\int_0^T \mathbf{u}_\perp^\wedge(t)\mathbf{u}_\perp(t)dt; U_0^2 = \frac{1}{T}\int_0^T \mathbf{u}_0^\wedge(t)\mathbf{u}_0(t)]dt = \frac{\mathbf{j}^\wedge\mathbf{j}}{9T}\int_0^T u_+^2(t)dt = \frac{1}{3T}\int_0^T u_+^2(t)dt$. Similarly, an arbitrary vector of line currents can be decomposed into orthogonal components:

$$\mathbf{i}(t) = \mathbf{i}_\perp(t) + \mathbf{i}_0(t), \tag{4.53}$$

then the integral power loss from eqn (4.50) is determined by the expression

$$\begin{aligned}
\Delta P &= \frac{1}{T}\int_0^T \mathbf{i}^\wedge(t)\mathbf{R}\mathbf{i}(t)dt = \frac{r}{T}\int_0^T \mathbf{i}^\wedge(t)[\mathbf{I} + (r_N/r)\mathbf{j}\mathbf{j}^\wedge]\mathbf{i}(t)dt = \\
&= \frac{r}{T}\int_0^T [\mathbf{i}_\perp(t)+\mathbf{i}_0(t)]^\wedge[\mathbf{i}_\perp(t)+\mathbf{i}_0(t)+\frac{3r_N}{r}\mathbf{i}_0(t)]dt = \\
&= \frac{r}{T}\int_0^T \mathbf{i}_\perp^\wedge(t)\mathbf{i}_\perp(t)dt + (1+\frac{3r_N}{r})\frac{r}{T}\int_0^T \mathbf{i}_0^\wedge(t)\mathbf{i}_0(t)]dt = \\
&= r\left(I_\perp^2 + \frac{r+3r_N}{r}I_0^2\right) = r\left(I_\perp^2 + \frac{1}{1-\sigma}I_0^2\right),
\end{aligned} \tag{4.54}$$

where $I_\perp^2 = \frac{1}{T}\int_0^T \mathbf{i}_\perp^\wedge(t)\mathbf{i}_\perp(t)dt; I_0^2 = \frac{1}{T}\int_0^T \mathbf{i}_0^\wedge(t)\mathbf{i}_0(t)]dt = \frac{1}{3T}\int_0^T i_+^2(t)dt.$

Substitution of eqn (4.52), (4.54) in eqn (4.48) gives a modification of the apparent power formula in the form [31]

$$S = \sqrt{[U_\perp^2 + (1-\sigma)U_0^2][I_\perp^2 + (1-\sigma)^{-1} I_0^2]}. \tag{4.55}$$

Unlike the Buchholz formula in the presence of zero-sequence components of phase voltages or line currents, the eqn (4.55) clearly shows the dependence of apparent power on the ratio of power cable resistances, which sets the coefficient σ.

The equal sign in the eqn (4.47), that is, a unit value of the integral power factor $\Lambda = P/S = 1$ takes place if the vectors $\mathbf{R}^{1/2}\mathbf{i}(t)$ and $\mathbf{R}^{-1/2}\mathbf{u}(t)$ are proportional. Therefore, the active current of the multiphase power system, which provides a given active load power *P* at minimal integral losses in the transmission line, should be sought in the form of

$$\mathbf{i}_A(t) = K\mathbf{R}^{-1}\mathbf{u}(t). \tag{4.56}$$

The value of the proportionality coefficient *K* is determined from the condition $\frac{1}{T}\int_0^T \mathbf{u}^\wedge(t)\mathbf{i}_A(t)dt = P$ as follows

$$K = \frac{P}{\frac{1}{T}\int_0^T \mathbf{u}^\wedge(t)\mathbf{R}^{-1}\mathbf{u}(t)dt}.$$

The substitution of this value in eqn (4.56) gives the final expression for the active current of the multiphase power system [33, 34]:

$$\mathbf{i}_A(t) = \frac{P}{\frac{1}{T}\int_0^T \mathbf{u}^\wedge(t)\mathbf{R}^{-1}\mathbf{u}(t)dt}\mathbf{R}^{-1}\mathbf{u}(t) = \frac{P}{P_0}\mathbf{i}_0(t). \tag{4.57}$$

Consequently, the active current of the multiphase power system becomes part of the short-circuit current vector that is equal to the ratio of load power to the short-circuit power of the voltages in resistances of the transmission line.

The mode of the multiphase power supply system with the minimum possible losses [33] $\Delta P_{\text{MIN}} = \Delta P_A = P^2/P_0$ caused by the active current (eqn (4.57)) corresponds to the maximum possible value of the integral power factor $\Lambda = P/\sqrt{\Delta P P_0}$ equal to unity. The energy-efficient strategy of active filtration proposed in [35] is to form an active current (eqn (4.57)) in the transmission line. Due to this the integral power losses gain is achieved [33]

$$W = \frac{\Delta P}{\Delta P_{\text{MIN}}} = \frac{\Delta P}{P^2/P_0} = \frac{S^2}{P^2} = \frac{1}{\Lambda^2}. \tag{4.58}$$

We will show that the known proportional-vector strategies for forming currents of a three-phase source with an integral value of the proportionality coefficient do not provide the unit value of the power factor [36].

Active filtration strategies that form an active current in the form of Fryze in the transmission line

$$\mathbf{i}_F(t) = \frac{P}{\frac{1}{T}\int_0^T \mathbf{u}^\wedge(t)\mathbf{u}(t)dt}\mathbf{u}(t) = \frac{\mathbf{u}\circ\mathbf{i}}{\mathbf{u}\circ\mathbf{u}}\mathbf{u}(t) = \frac{P}{U^2}\mathbf{u}(t), \tag{4.59}$$

have a square of apparent power following eqn (4.55)

$$\begin{aligned} S^2 &= [U_\perp^2+(1-\sigma)U_0^2][I_\perp^2+(1-\sigma)^{-1}I_0^2] = \\ &= [U_\perp^2+(1-\sigma)U_0^2]\left(\frac{P}{U^2}\right)^2[U_\perp^2+U_0^2/(1-\sigma)] = \\ &= P^2\frac{(U^2-\sigma U_0^2)\{U^2+[(1-\sigma)^{-1}-1]U_0^2\}}{U^4} \\ &= P^2(1-\sigma\Delta_0^2)[1+\sigma(1-\sigma)^{-1}\Delta_0^2], \end{aligned} \tag{4.60}$$

where $\Delta_0^2 = U_0^2/U^2$ is the integral voltage zero-sequence factor.

Further transformations of the expression represented by eqn (4.60) result in:

$$\begin{aligned} S^2 &= P^2[1-\sigma\Delta_0^2+\sigma(1-\sigma)^{-1}\Delta_0^2-\sigma^2(1-\sigma)^{-1}\Delta_0^4] = \\ &= P^2[1-\sigma(1-\sigma)^{-1}(1-\sigma)\Delta_0^2+\sigma(1-\sigma)^{-1}\Delta_0^2-\sigma^2(1-\sigma)^{-1}\Delta_0^4] = \\ &= P^2[1+\sigma^2(1-\sigma)^{-1}\Delta_0^2-\sigma^2(1-\sigma)^{-1}\Delta_0^4] = \\ &P^2[1+\sigma^2(1-\sigma)^{-1}\Delta_0^2(1-\Delta_0^2)]. \end{aligned} \tag{4.61}$$

From eqn (4.61), it follows that the power factor in the case of using the S. Fryze strategy, provided $\Delta_0^2 \neq 0$ does not reach a unit value:

$$\Lambda_F = \left[1+\frac{\sigma^2}{1-\sigma}\Delta_0^2(1-\Delta_0^2)\right]^{-1/2} < 1. \tag{4.62}$$

Similarly, the power factor does not reach a unit value in the case of neutral current compensation by the proportional-vector formation of phase currents with a reference vector of phase voltages utterly devoid of the zero-sequence component, according to the algorithm by P. Salmeron and J. Montano [37]. In this case, the current vector of the three-phase source

$$\mathbf{i}_S(t) = \frac{P}{U_\perp^2}\mathbf{u}_\perp(t); \tag{4.63}$$

apparent power by eqn (4.55)

$$S^2 = [U_\perp^2 + (1-\sigma)U_0^2]\left(\frac{P}{U_\perp^2}\right)^2 U_\perp^2 =$$
$$= P^2[1 + (1-\sigma)U_0^2/U_\perp^2] = P^2[1 + (1-\sigma)\Delta_0^2/(1-\Delta_0^2)]; \qquad (4.64)$$

and the power factor

$$\Lambda_S = \left[1 + (1-\sigma)\frac{\Delta_0^2}{1-\Delta_0^2}\right]^{-1/2} < 1. \qquad (4.65)$$

Graphs constructed according to eqn (4.62) and (4.65) at different coefficient values σ are shown in Figure 4.5 and Figure 4.6, respectively.

They illustrate the advantages on the integral power factor of the proposed SAF control strategy with an optimally weakened zero-sequence component compared to the S. Fryze algorithm with apparent preservation of the zero-sequence component (according to Figure 4.5 up to 7%) and compared to the P. Salmeron, J. Montano algorithm with complete suppression of the voltage zero-sequence component (according to Figure 4.6 up to 1.5%) [36].

Let us compare the proposed formulas for determining apparent power (eqn (4.48), (4.50), (4.55)) with standardized values, implying that there are American and European approaches to determining apparent power [38].

The American approach, summarized in IEEE Std 1459-2010 [39], developed by a team of scientists led by Emanuel A.E., offers a formula for determining apparent power:

$$S_e = 3U_e I_e, \qquad (4.66)$$

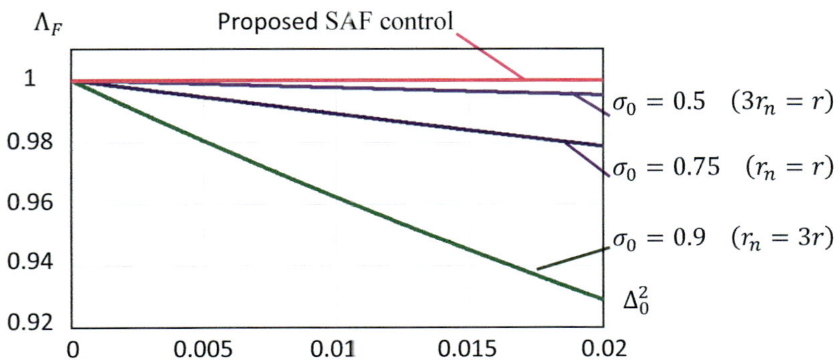

Figure 4.5 Dependence of the power factor on the integral voltage zero-sequence factor when applying the S. Fryze active filtering strategy.

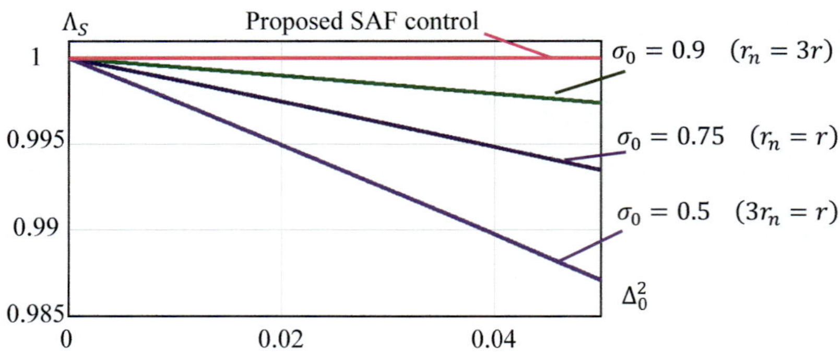

Figure 4.6 Dependence of the power factor on the integral voltage zero-sequence factor when applying the P. Salmeron and J. Montano active filtering strategy.

where $U_e = \sqrt{\frac{3(U_A^2+U_B^2+U_C^2)+\xi(U_{AB}^2+U_{BC}^2+U_{CA}^2)}{9(1+\xi)}}$ is effective phase voltage, $I_e = \sqrt{\frac{I_A^2+I_B^2+I_C^2+(r_N/r)I_N^2}{3}}$ is equivalent current, expressed through the squares of phase and line voltages values, line current and neutral current, respectively; $\xi = P_\Delta/P_Y = 3R_Y/R_\Delta$ is a coefficient equal to the ratio of active powers due to the specified phase voltages of the four-wire power system in parts of the conditional load connected by a triangle and a star (Figure 4.7).

This definition of apparent power leads to some difficulties. Firstly, the ratio of resistances R_Y/R_Δ and the value of the coefficient, in general, are unknown. Then it is recommended [39] to take $\xi = 1$, what corresponds to the unreasonable equality of these active powers $P_Y = P_\Delta$. Secondly, the change in the ratio between the active resistances r_N/r only impacts the equivalent current, while having no influence on the effective phase voltage, which leads to certain inconsistencies [38].

For greater clarity, we convert the components of the expression given in eqn (4.66), taking into account the previously entered designations as follows:

$$I_A^2+I_B^2+I_C^2+(r_N/r)I_N^2 = \Delta P/r = \frac{1}{Tr}\int_0^T \mathbf{i}^\wedge(t)\overline{\mathbf{R}}\mathbf{i}(t)dt$$
$$= \left(I_\perp^2+\frac{1}{1-\sigma}I_0^2\right); U_A^2+U_B^2+U_C^2 = \frac{1}{T}\int_0^T \mathbf{u}^\wedge(t)\mathbf{u}(t)dt =$$
$$= \frac{1}{T}\left[\int_0^T \mathbf{u}_\perp^\wedge(t)\mathbf{u}_\perp(t)dt+\int_0^T \mathbf{u}_0^\wedge(t)\mathbf{u}_0(t)]dt\right] = U_\perp^2+U_0^2;$$

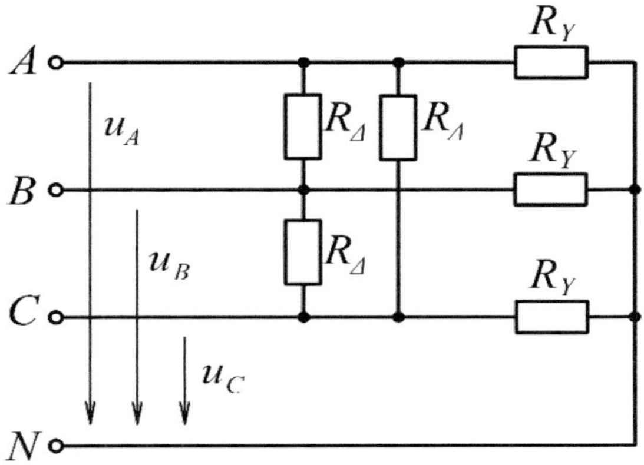

Figure 4.7 Standardized model of three-phase four-wire power supply system.

$$U_{AB}^2+U_{BC}^2+U_{CA}^2=\frac{1}{T}\int_0^T\left\|\begin{matrix}u_{AB}(t) & u_{BC}(t) & u_{CA}(t)\end{matrix}\right\|\left\|\begin{matrix}u_{AB}(t)\\ u_{BC}(t)\\ u_{CA}(t)\end{matrix}\right\|dt=$$

$$=\frac{1}{T}\int_0^T\mathbf{u}^\wedge(t)\left\|\begin{matrix}1 & 0 & -1\\ -1 & 1 & 0\\ 0 & -1 & 1\end{matrix}\right\|\left\|\begin{matrix}1 & -1 & 0\\ 0 & 1 & -1\\ -1 & 0 & 1\end{matrix}\right\|\mathbf{u}(t)dt=$$

$$=\frac{1}{T}\int_0^T\mathbf{u}^\wedge(t)\left\|\begin{matrix}2 & -1 & -1\\ -1 & 2 & -1\\ -1 & -1 & 2\end{matrix}\right\|\mathbf{u}(t)dt=$$

$$=\frac{1}{T}\int_0^T[\mathbf{u}_\perp(t)+\mathbf{u}_0(t)]^\wedge(3\mathbf{I}-\mathbf{jj}^\wedge)[\mathbf{u}_\perp(t)+\mathbf{u}_0(t)]dt$$

$$=\frac{3}{T}\int_0^T\mathbf{u}_\perp^\wedge(t)\mathbf{u}_\perp dt=3U_\perp^2.$$

Let us establish a relationship between the values of the resistances R_Y, R_Δ and r_N, r in which the eqn (4.55) and (4.66) are equivalent. By squaring both parts of eqn (4.66) considering the designations of eqn (4.55) and the identity of the second current multiplier in both expressions, we will have:

$$S_e^2=\frac{(U_A^2+U_B^2+U_C^2)+\xi(U_{AB}^2+U_{BC}^2+U_{CA}^2)/3}{(1+\xi)}\times[I_\perp^2+(1-\sigma)^{-1}I_0^2]. \tag{4.67}$$

Voltage multiplier from eqn (4.66) is converted to

$$\frac{(U_A^2+U_B^2+U_C^2)+\xi(U_{AB}^2+U_{BC}^2+U_{CA}^2)/3}{(1+\xi)}=\frac{(U_\perp^2+U_0^2)+\xi U_\perp^2}{(1+\xi)}=U_\perp^2+\frac{U_0^2}{1+\xi}.$$

By comparing it with the first multiplier of the eqn (4.55), we set the equivalence condition

$$1-\sigma = \frac{1}{1+\xi},$$

where, taking into account the values of the coefficients σ and the ξ, the equivalence ratio follows

$$R_{\mathrm{Y}}/R_{\Delta} = r_N/r. \tag{4.68}$$

Consequently, the proposed formulas of apparent power given by eqn (4.48) and (4.55) are entirely identical to IEEE standard [39] if you choose the resistance parameters of model (Figure 4.7) in order to calculate the effective voltage of the resulting equivalent ratio eqn (4.68)). The equivalence ratio (eqn (4.68)) eliminates the uncertainty of the coefficient ξ in the specified standard. Let us consider the limiting cases of right side (eqn (4.68)). The case $r_N/r{\rightarrow}\infty$ corresponds $R_{\mathrm{Y}}{\rightarrow}\infty$, that is, in the scheme in Figure 4.7; there are only resistors connected to triangular; if $r_N/r{\rightarrow}0$, $R_{\Delta}{\rightarrow}\infty$ and in the diagram in Figure 4.7, there are only resistors connected by a star [45].

The European approach for the apparent power definition is based on the M. Depenbrock concept, considering a three-phase zero-wire power system as equivalent to a four-phase one. This approach is based on the DIN std. 40110, 2002 [40] and is illustrated in detail in [38] for the sinusoidal mode of a three-phase power system with different values of line and neutral wire resistances. Apparent power is represented by the product of norms of four-dimensional complex vectors of current and voltage:

$$S = |\dot{\mathbf{v}}||\dot{\mathbf{i}}| \tag{4.69}$$

where

$$\dot{\mathbf{v}} = \left\| \frac{\dot{V}_A-\dot{V}_{\mathrm{REF}}}{\sqrt{\rho_A}} \frac{\dot{V}_B-\dot{V}_{\mathrm{REF}}}{\sqrt{\rho_B}} \frac{\dot{V}_C-\dot{V}_{\mathrm{REF}}}{\sqrt{\rho_C}} \frac{\dot{V}_N-\dot{V}_{\mathrm{REF}}}{\sqrt{\rho_N}} \right\|^{\wedge};$$

$$\dot{\mathbf{i}} = \left\| \dot{I}_A\sqrt{\rho_A}\dot{I}_B\sqrt{\rho_B}\dot{I}_C\sqrt{\rho_C}\dot{I}_N\sqrt{\rho_N} \right\|^{\wedge};$$

$\dot{V}_{\mathrm{REF}} = \frac{\sum_{\forall k}\dot{V}_k/\rho_k}{\sum_{\forall k}1/\rho_k}$;$k = A, B, C, N$;$\rho_k = r_k/r$;$r$ is the essential parameter of power cable resistances; $\dot{V}_A, \dot{V}_B, \dot{V}_C, \dot{V}_N$;$\dot{I}_A, \dot{I}_B, \dot{I}_C, \dot{I}_N$ are complexes of operating voltage and current values of an equivalent four-phase power supply system in sinusoidal mode. The disadvantages of the approach are the increased dimension of voltage and current vectors describing the three-phase power system; the opaque physical content of a complex quantity $\dot{V}_{\mathrm{REF}}$; consideration of the sinusoidal mode of the power system only [45].

Within the framework of the proposed approaches for determining the apparent power, we consider a comprehensive scheme for replacing the power system for the short-circuit mode of a three-phase source (Figure 4.8.), where the complex potentials of the nodes are indicated as $\dot{V}_A, \dot{V}_B, \dot{V}_C, \dot{V}_N$.

We denote $\dot{V}_0$ the complex potential of the resistors common point and find its value from the equation drawn up according to Kirchhoff's first law:

$$\frac{\dot{V}_A-\dot{V}_0}{r_A}+\frac{\dot{V}_B-\dot{V}_0}{r_B}+\frac{\dot{V}_C-\dot{V}_0}{r_C}=\frac{\dot{V}_0-\dot{V}_N}{r_N},$$

where

$$\dot{V}_0=\frac{\dot{V}_A r_A^{-1}+\dot{V}_B r_B^{-1}+\dot{V}_C r_C^{-1}+\dot{V}_N r_N^{-1}}{r_A^{-1}+r_B^{-1}+r_C^{-1}+r_N^{-1}}=\frac{\sum_{\forall k}\dot{V}_k/\rho_k}{\sum_{\forall k}1/\rho_k}=\dot{V}_{\mathrm{REF}}.$$

Consequently, the value $\dot{V}_{\mathrm{REF}}$ from [38] is the complex potential of the resistors common point of the power system replacement scheme in the short-circuit mode of the three-phase source. Then the voltage multiplier square of apparent power by the eqn (4.69)

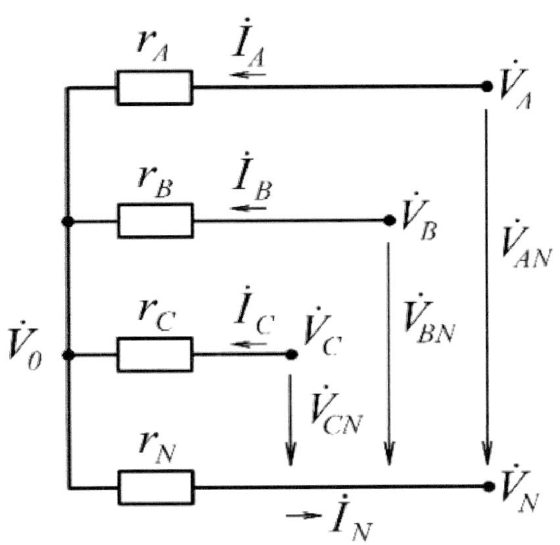

Figure 4.8 Resistive model of three-phase four-wire power supply system in sinusoidal mode.

$$|\dot{\mathbf{v}}|^2 = \frac{\left|\dot{V}_A - \dot{V}_{REF}\right|^2}{\rho_A} + \frac{\left|\dot{V}_B - \dot{V}_{REF}\right|^2}{\rho_B} + \frac{\left|\dot{V}_C - \dot{V}_{REF}\right|^2}{\rho_C} + \frac{\left|\dot{V}_N - \dot{V}_{REF}\right|^2}{\rho_N} =$$

$$= \left(\frac{\left|\dot{V}_A - \dot{V}_0\right|^2 r}{r_A} + \frac{\left|\dot{V}_B - \dot{V}_0\right|^2 r}{r_B} + \frac{\left|\dot{V}_C - \dot{V}_0\right|^2 r}{r_C} + \frac{\left|\dot{V}_0 - \dot{V}_N\right|^2 r}{r_N} \right) = P_0 r$$

is the normalized short-circuit power of power supply system in Figure 4.8.

Let us show that the voltage integral from eqn (4.48) gives the power of losses P_0 of the power system short circuit for the case of various resistances of the line wires under consideration.

By sinusoidal forms of phase voltages

$$\frac{1}{T}\int_0^T \mathbf{u}^{\wedge}(t)\mathbf{R}^{-1}\mathbf{u}(t)dt = \left\| \begin{matrix} \dot{V}_A - \dot{V}_N \\ \dot{V}_B - \dot{V}_N \\ \dot{V}_C - \dot{V}_N \end{matrix} \right\|^{\wedge} \mathbf{R}^{-1} \left\| \begin{matrix} \dot{V}_A - \dot{V}_N \\ \dot{V}_B - \dot{V}_N \\ \dot{V}_C - \dot{V}_N \end{matrix} \right\|^{*}, \tag{4.70}$$

where * is the sign of complex conjugating. Considering the value of the inverse loss matrix for different resistances of line wires

$$\mathbf{R}^{-1} = \left(\left\| \begin{matrix} r_A & 0 & 0 \\ 0 & r_B & 0 \\ 0 & 0 & r_C \end{matrix} \right\| + r_N \left\| \begin{matrix} 1 \\ 1 \\ 1 \end{matrix} \right\| \| \; 1 \;\; 1 \;\; 1 \; \| \right)^{-1} =$$

$$= \left\| \begin{matrix} r_A^{-1} & 0 & 0 \\ 0 & r_B^{-1} & 0 \\ 0 & 0 & r_C^{-1} \end{matrix} \right\| - \frac{1}{r_A^{-1} + r_B^{-1} + r_C^{-1} + r_N^{-1}} \left\| \begin{matrix} r_A^{-1} \\ r_B^{-1} \\ r_C^{-1} \end{matrix} \right\| \| \; r_A^{-1} \;\; r_B^{-1} \;\; r_C^{-1} \; \|,$$

eqn (4.70) can be represented as:

$$\frac{1}{T}\int_0^T \mathbf{u}^{\wedge}(t)\mathbf{R}^{-1}\mathbf{u}(t)dt = \left\| \dot{V}_A - \dot{V}_N \dot{V}_B - \dot{V}_N \dot{V}_C - \dot{V}_N \right\| \times$$

$$\times \left(\left\| \begin{matrix} r_A^{-1} & 0 & 0 \\ 0 & r_B^{-1} & 0 \\ 0 & 0 & r_C^{-1} \end{matrix} \right\| - \frac{1}{r_A^{-1} + r_B^{-1} + r_C^{-1} + r_N^{-1}} \left\| \begin{matrix} r_A^{-1} \\ r_B^{-1} \\ r_C^{-1} \end{matrix} \right\| \|r_A^{-1} \; r_B^{-1} \; r_C^{-1}\| \right) \times$$

$$\times \left\| \begin{matrix} \dot{V}_A - \dot{V}_N \\ \dot{V}_B - \dot{V}_N \\ \dot{V}_C - \dot{V}_N \end{matrix} \right\| = \left\| \left(\dot{V}_A - \dot{V}_0\right) r_A^{-1} \left(\dot{V}_B - \dot{V}_0\right) r_B^{-1} \left(\dot{V}_C - \dot{V}_0\right) r_C^{-1} \right\| \times$$

$$\times \left(\left\| \begin{matrix} \dot{V}_A - \dot{V}_0 \\ \dot{V}_B - \dot{V}_0 \\ \dot{V}_C - \dot{V}_0 \end{matrix} \right\|^{*} + \left\| \begin{matrix} \dot{V}_0 - \dot{V}_N \\ \dot{V}_0 - \dot{V}_N \\ \dot{V}_0 - \dot{V}_N \end{matrix} \right\|^{*} \right) =$$

$$
= \frac{\left|\dot{V}_A - \dot{V}_0\right|^2}{r_A} + \frac{\left|\dot{V}_B - \dot{V}_0\right|^2}{r_B} + \frac{\left|\dot{V}_C - \dot{V}_0\right|^2}{r_C} +
$$

$$
+ \left(\dot{V}_0 - \dot{V}_N\right)^* \left(\frac{\dot{V}_A - \dot{V}_0}{r_A} + \frac{\dot{V}_B - \dot{V}_0}{r_B} + \frac{\dot{V}_C - \dot{V}_0}{r_C}\right) =
$$

$$
= \frac{\left|\dot{V}_A - \dot{V}_0\right|^2}{r_A} + \frac{\left|\dot{V}_B - \dot{V}_0\right|^2}{r_B} + \frac{\left|\dot{V}_C - \dot{V}_0\right|^2}{r_C} + \frac{\left|\dot{V}_0 - \dot{V}_N\right|^2}{r_N} = \frac{|\dot{\mathbf{v}}|^2}{r} = P_0.
$$

Similarly, we show that the square of the current multiplier of the apparent power of eqn (4.69) is the normalized power of losses of the three-phase power system determined by the current integral from eqn (4.48):

$$
\left|\dot{\mathbf{i}}\right|^2 = I_A^2\rho_A + I_B^2\rho_B + I_C^2\rho_C + I_N^2\rho_N = r^{-1}[\dot{I}_A r_A (\dot{I}_A)^* + \dot{I}_B r_B (\dot{I}_B)^*
$$

$$
+ \dot{I}_C r_C (\dot{I}_C)^* + + (\dot{I}_A + \dot{I}_B + \dot{I}_C) r_N (\dot{I}_A + \dot{I}_B + \dot{I}_C)^* =
$$

$$
= r^{-1} \left\| \dot{I}_A \ \dot{I}_B \ \dot{I}_C \right\| \left(\left\| \begin{matrix} r_A & 0 & 0 \\ 0 & r_B & 0 \\ 0 & 0 & r_C \end{matrix} \right\| + r_N \left\| \begin{matrix} 1 \\ 1 \\ 1 \end{matrix} \right\| \left\| 1 \ 1 \ 1 \right\| \right) \left\| \begin{matrix} \dot{I}_A \\ \dot{I}_B \\ \dot{I}_C \end{matrix} \right\|^* =
$$

$$
= r^{-1} \frac{1}{T} \int_0^T \mathbf{i}^\wedge(t) \mathbf{R} \mathbf{i}(t) dt = r^{-1} \Delta \mathrm{P}.
$$

Thus, the equivalence of the proposed formulas represented by eqn (4.48) and (4.50) to the European approach of determining apparent power in the sinusoidal mode for the case of different values of the transmission line resistances is shown.

In three-phase four-wire power systems, the asymmetry of phase voltages and line currents leads to increased energy losses due to the flow of the total current of line wires in neutral, which adequately reflects the formula for determining apparent power (eqn (4.48)), which takes into account the parameters of the power cable resistances. In this section, the determination of the components of apparent power and corresponding normalized currents is brought in line with the new formula of apparent power, and the average quadratic value of each normalized current corresponds to its contribution to the total power of losses.

The expression for the active current contains a zero-sequence component and, in general, higher harmonics, which also transfer part of the energy to the load, but at the same time, undesirable components of the current are generated into the network, which worsens the Total Harmonic Distortion (THD) indicator. Therefore, next, we will consider the options for decomposition of apparent power into quadratic components, reflected in international standards, and the corresponding strategies for energy-efficient control of active filtration tools [41–45].

The IEEE 1459-2010 standard [39] generalizes the apparent power decomposition. Each of its quadratic components corresponds to the current decomposition into orthogonal components and the selection of individual components for the control strategy construction of energy-efficient active filtering. The advantages and disadvantages of each strategy further are considered.

Apparent power schedule for active and inactive power [39] is:

$$S^2 = P^2 + N^2. \tag{4.71}$$

It corresponds to the use of the entire spectral composition of the phase voltage vector $\mathbf{u}(t)$ for the transfer of energy to the load, and the strategy of shunt active filtration consists in the proportional-vector formation of the current vector of the three-phase source:

$$\mathbf{i}_\sigma(t) = \mathbf{i}_A(t) = G_\sigma r\mathbf{R}^{-1}\mathbf{u}(t) = G_\sigma[\mathbf{u}_\perp(t) + (1-\sigma)\mathbf{u}_0(t)] = G_\sigma \mathbf{u}_\sigma(t), \tag{4.72}$$

where $G_\sigma = \frac{P}{U_\perp^2+(1-\sigma)U_0^2}$ is the proportionality coefficient, the value calculated from the condition of zero energy consumption by the active filter for the switching period [26]. At the same time, we have the minimum power loss

$$\Delta P_{\mathrm{MIN}} = \frac{1}{T}\int_0^T \mathbf{i}_\sigma^\wedge(t)\mathbf{R}\mathbf{i}_\sigma(t)dt = \frac{G_\sigma^2}{T}\int_0^T \mathbf{u}_\sigma^\wedge(t)\mathbf{R}\mathbf{u}_\sigma(t)dt =$$
$$= \frac{G_\sigma^2 r^2}{T}\int_0^T \mathbf{u}^\wedge(t)\mathbf{R}^{-1}\mathbf{u}(t)dt = \frac{P^2 r}{U_\perp^2+(1-\sigma)U_0^2} = rG_\sigma P,$$

that provides the unity power factor.

The disadvantage of the decomposition (eqn (4.74)) from a theoretical point of view, according to Budeanu, is the presence in the inactive component of reactive power in the form of an algebraic sum of harmonic reactive power, which is not a good indicator of the heat load on the transmission line [46]. Furthermore, to properly account for the action of higher-order harmonic currents in transmission lines, it is necessary to consider skin effects [39], which significantly complicate the calculations. In practical terms, the effective load current according to eqn (4.72) contains a zero-sequence component and is generally high harmonic, which also transfers some of the energy to the load, but at the same time the undesirable harmonic component of the current is generated in the network, which worsens the THD index.

Decomposition of apparent power by the apparent power distribution of the first harmonics is

$$S^2 = S_1^2 + S_N^2,$$

and it corresponds to the energy transfer at the main frequency. The energy-efficient SAF control strategy provides in this case the allocation of the phase voltage vector of main harmonic $\mathbf{u}_1(t)$ and the proportional-vector formation of the source current vector [47]

$$\begin{aligned}\mathbf{i}_{1\sigma}(t) &= G_{1\sigma} r \mathbf{R}^{-1}\mathbf{u}_1(t) = G_{1\sigma}\left[\mathbf{u}_{1+}(t) + \mathbf{u}_{1-}(t) + (1-\sigma)\,\mathbf{u}_{10}(t)\right] = \\ &= G_{1\sigma}\mathbf{u}_{1\sigma}(t),\end{aligned}$$

where $\mathbf{u}_{1+}(t)$,$\mathbf{u}_{1-}(t)$,$\mathbf{u}_{10}(t)$ are the positive, negative, and zero-sequence components of the phase voltage vector of main harmonic; $G_{1\sigma} = \frac{P}{\frac{1}{T}\int_0^T \mathbf{u}_{1\sigma}^{\wedge}(t)\mathbf{u}_{1\sigma}(t)dt} = \frac{P}{U_{1+}^2 + U_{1-}^2 + U_{10}^2(1-\sigma)}$ is proportionality coefficient that provides zero energy consumption by an active filter for the switching period; U_{1+}, U_{1-}, U_{10} are the r.m.s. values of the corresponding components.

The following expression determines power losses in this case

$$\begin{aligned}\Delta P_{1\sigma} &= \frac{1}{T}\int_0^T \mathbf{i}_{1\sigma}^{\wedge}(t)\mathbf{R}\mathbf{i}_{1\sigma}(t)dt = \frac{G_{1\sigma}^2 r^2}{T}\int_0^T \mathbf{u}_1^{\wedge}(t)\mathbf{R}^{-1}\mathbf{u}_1(t)dt = \\ &= \frac{P^2 r}{U_{1+}^2 + U_{1-}^2 + U_{10}^2(1-\sigma_0)} = rG_{1\sigma}P.\end{aligned}$$

The integral power losses gain in the case of using this control strategy is determined as follows

$$\begin{aligned}W_{1\sigma} = \frac{\Delta P}{\Delta P_{1\sigma}} &= \frac{S^2 r}{U_\perp^2 + (1-\sigma)U_0^2} \div \frac{P^2 r}{U_{1+}^2 + U_{1-}^2 + U_{10}^2(1-\sigma)} \\ &= \frac{U_{1+}^2 + U_{1-}^2 + U_{10}^2(1-\sigma)}{[U_\perp^2 + (1-\sigma)U_0^2]\Lambda^2},\end{aligned}$$

and is inferior to the first control strategy in the presence of voltage higher harmonics. The use of this SAF control strategy is also accompanied by the emission into the network of components of the reverse and zero sequences of the main frequency.

Ultimately, the final analysis separates the apparent power into the following squared component terms:

- active power of the component of the positive sequence of the main frequency P_{1+};
- reactive power of the component of the positive sequence of the main frequency Q_{1+};
- unbalance power $S_U = \sqrt{S^2 - P_{1+}^2 - Q_{1+}^2}$.

Following this decomposition, the transfer of energy to the load is carried out at the frequency of the main harmonic by vector of the voltage-positive sequence component $\mathbf{u}_{1+}(t)$. The current vector of the source is formed as follows

$$\mathbf{i}_{1+}(t) = \frac{P}{\frac{1}{T}\int_0^T [\mathbf{u}_{1+}(t)]^\wedge \mathbf{u}_{1+}(t)dt}\mathbf{u}_{1+}(t) = \frac{P}{U_{1+}^2}\mathbf{u}_{1+}(t).$$

The power losses gain with this SAF control strategy

$$W_{1+} = \frac{\Delta P}{\Delta P_{1+}} = \frac{S^2 r}{U_\perp^2 + (1-\sigma)U_0^2} \div \frac{1}{T}\int_0^T [\mathbf{i}_{1+}(t)]^\wedge \mathbf{R}\mathbf{i}_{1+}(t)dt =$$

$$= \frac{S^2 r}{U_\perp^2 + (1-\sigma)U_0^2} \div \frac{rP^2}{U_{1+}^2} = \frac{U_{1+}^2}{[U_\perp^2 + U_0^2(1-\sigma)]\Lambda^2}.$$

As a result, a high-quality spectral configuration of the consumed current can be obtained by using the voltage-positive sequence component of the main harmonic as the reference vector for the effective current formation, but such a strategy of SAF control has inferior power losses than a strategy using a phase voltage vector with optimally weakened zero phase components.

4.4 Assessment and Improvement of Electric Energy Quality of Distributed Generation Network Systems

Unequal loading across the phases of a multiphase system leads to degraded power quality, manifesting as additional currents in the transmission line that do not transfer active or reactive power but distort the voltage at common coupling points [48]. In such cases, the additional energy consumption by the supplier due to the unbalanced load is not paid for by the consumer. The quantitative measure of load asymmetry is unbalanced power, but the actual standard [49] and calculation method [50] only define the square of this value, complicating hardware compensation, measurement, and accounting of unbalanced power. Control strategies for shunt active filters (SAFs) to compensate source currents associated with unbalanced power are based on several power theories [51–55], which link unbalanced power to a specific component of the load current vector in different ways. This connection is most transparent for linear loads within the currents' physical component theory [56], where unbalanced power is associated with the current component proportional to the main voltage negative sequence component. However, the proportionality coefficient is determined by the complex conductivity of individual phase loads, complicating the application of this theory for filtering alternating load currents. The main idea behind distributed active filtering involves dividing the task of compensating higher harmonics and unwanted power components among semiconductor inverters of renewable sources [57–59]. In this regard, for convenient control of unbalanced power, it is advisable to present it as orthogonal components that contribute separately to the apparent power [60].

The steady-state process in a three-phase, three-wire power system with a symmetrical sinusoidal power supply and linear loads is determined by the three-dimensional vectors of phase voltages and line currents shown in the time domain

$$\mathbf{u}(t) = \begin{Vmatrix} u_A(t) \\ u_B(t) \\ u_C(t) \end{Vmatrix} = \sqrt{2} \begin{Vmatrix} U_\Phi \cos(\omega t) \\ U_\Phi \cos(\omega t - 2\pi/3) \\ U_\Phi \cos(\omega t + 2\pi/3) \end{Vmatrix};$$

$$\mathbf{i}(t) = \begin{Vmatrix} i_A(t) \\ i_B(t) \\ i_C(t) \end{Vmatrix} = \sqrt{2} \begin{Vmatrix} I_A \cos(\omega t + \psi_A) \\ I_B \cos(\omega t + \psi_B) \\ I_C \cos(\omega t + \psi_C) \end{Vmatrix},$$

where the U_Φ is phase voltage RMS value; $\omega = 2\pi/T$ is the main frequency; I_A, I_B, I_C are RMS values of the line currents with corresponding initial phases ψ_A, ψ_B, and ψ_C These vectors have the following complex form in the frequency domain that may be disseminate with the unit vectors of positive and negative sequences components:

$$\mathbf{u}(t) \leftrightarrow \overline{\mathbf{u}} = \begin{Vmatrix} \dot{U}_A \\ \dot{U}_B \\ \dot{U}_C \end{Vmatrix} = U_\Phi \begin{Vmatrix} 1 \\ e^{-j2\pi/3} \\ e^{j2\pi/3} \end{Vmatrix} = \frac{U}{\sqrt{3}} \begin{Vmatrix} 1 \\ \tilde{a} \\ \dot{a} \end{Vmatrix} = U\overline{\mathbf{e}}_+;$$

$$\mathbf{i}(t) \leftrightarrow \overline{\mathbf{i}} = \begin{Vmatrix} \dot{I}_A \\ \dot{I}_B \\ \dot{I}_C \end{Vmatrix} = \begin{Vmatrix} I_A e^{j\psi_A} \\ I_B e^{j\psi_B} \\ I_C e^{j\psi_C} \end{Vmatrix} = \dot{I}_+ \overline{\mathbf{e}}_+ + \dot{I}_- \overline{\mathbf{e}}_-, \tag{4.73}$$

where $U = U_\Phi\sqrt{3}$ is RMS value of line voltage; $\overline{\mathbf{e}}_+$ is the unit vector of positive sequence; $\overline{\mathbf{e}}_- = \|1\dot{a}\tilde{a}\|^\wedge/\sqrt{3} = \overline{\mathbf{e}}_+^*$ is the unit vector of negative sequence; $\dot{a} = e^{j2\pi/3}$; $\tilde{a} = e^{-j2\pi/3}$; $\dot{I}_+ = I_{R+} + jI_{I+}$; $\dot{I}_- = I_{R-} + jI_{I-}$ are complex coefficients depending on a particular type of linear load [60]. For its definition, we find the scalar product of the comprehensive vector of currents and the unit vector of symmetric components:

$$\begin{aligned} \dot{I}_+ &= \bar{\mathbf{i}}^{\wedge}\bar{\mathbf{e}}_- = \bar{\mathbf{i}}^{\wedge}\bar{\mathbf{u}}^*/U = (\bar{\mathbf{u}}^{\wedge}\bar{\mathbf{i}}^*)^*/U = (P - jQ)/U; \\ \dot{I}_- &= \bar{\mathbf{i}}^{\wedge}\bar{\mathbf{e}}_+ = \bar{\mathbf{i}}^{\wedge}\bar{\mathbf{u}}_-^*/U = (D_R - jD_I)/U. \end{aligned} \tag{4.74}$$

In the obtained expressions $\bar{\mathbf{u}}_- = \bar{\mathbf{u}}^* = \left\| \dot{U}_A \quad \dot{U}_C \quad \dot{U}_B \right\|$; *P*, *Q* are active and reactive power, D_R, D_I are scalar coefficients of the unbalanced power decomposition on orthogonal components [60]. The values of the introduced scalar coefficients can be obtained from eqn (4.74)

$$I_{R+} = \frac{P}{U}; I_{I+} = -\frac{Q}{U}; I_{R-} = \frac{D_R}{U}; I_{I-} = -\frac{D_I}{U},$$

and to present the current vector decomposition by scalar power components and reference voltage vectors:

$$\begin{aligned} \bar{\mathbf{i}} &= \frac{P-jQ}{U}\bar{\mathbf{e}}_+ + \frac{D_R-jD_I}{U}\bar{\mathbf{e}}_- = \frac{P-jQ}{U^2}\bar{\mathbf{u}} + \frac{D_R-jD_I}{U}\bar{\mathbf{u}}^* = \\ &= \frac{P}{U^2}\left\| \begin{matrix} \dot{U}_A \\ \dot{U}_B \\ \dot{U}_C \end{matrix} \right\| + \frac{Q}{U^2}\left\| \begin{matrix} -j\dot{U}_A \\ -j\dot{U}_B \\ -j\dot{U}_C \end{matrix} \right\| + \frac{D_R}{U^2}\left\| \begin{matrix} \dot{U}_A \\ \dot{U}_C \\ \dot{U}_B \end{matrix} \right\| + \frac{D_I}{U^2}\left\| \begin{matrix} -j\dot{U}_A \\ -j\dot{U}_C \\ -j\dot{U}_B \end{matrix} \right\| = \\ &= \frac{P\bar{\mathbf{u}}_P + Q\bar{\mathbf{u}}_Q + D_R\bar{\mathbf{u}}_{DR} + D_I\bar{\mathbf{u}}_{DI}}{U^2} = \bar{\mathbf{i}}_P + \bar{\mathbf{i}}_Q + \bar{\mathbf{i}}_{DR} + \bar{\mathbf{i}}_{DI}. \end{aligned} \tag{4.75}$$

It is easy to verify that the complex reference vectors of all the voltages obtained are mutually orthogonal, so that the apparent power according to Buchholz can be properly decomposed on quadratic components:

$$S^2 = U^2(\bar{\mathbf{i}}^{\wedge}\bar{\mathbf{i}}^*) = P^2 + Q^2 + D_R^2 + D_I^2. \tag{4.76}$$

The complex reference voltage vectors can also be represented in terms of the line-to-line voltage values

$$\begin{aligned} \bar{\mathbf{u}}_P &= \left\| \begin{matrix} \dot{U}_A \\ \dot{U}_B \\ \dot{U}_C \end{matrix} \right\| = \frac{1}{3}\left\| \begin{matrix} \dot{U}_{AB} - \dot{U}_{CA} \\ \dot{U}_{BC} - \dot{U}_{AB} \\ \dot{U}_{CA} - \dot{U}_{BC} \end{matrix} \right\|; \bar{\mathbf{u}}_Q = \left\| \begin{matrix} -j\dot{U}_A \\ -j\dot{U}_B \\ -j\dot{U}_C \end{matrix} \right\| = \frac{1}{\sqrt{3}}\left\| \begin{matrix} \dot{U}_{BC} \\ \dot{U}_{CA} \\ \dot{U}_{AB} \end{matrix} \right\|; \\ \bar{\mathbf{u}}_{DR} &= \left\| \begin{matrix} \dot{U}_A \\ \dot{U}_C \\ \dot{U}_B \end{matrix} \right\| = \frac{1}{3}\left\| \begin{matrix} \dot{U}_{AB} - \dot{U}_{CA} \\ \dot{U}_{CA} - \dot{U}_{BC} \\ \dot{U}_{BC} - \dot{U}_{AB} \end{matrix} \right\|; \bar{\mathbf{u}}_{DI} = \left\| \begin{matrix} -j\dot{U}_A \\ -j\dot{U}_C \\ -j\dot{U}_B \end{matrix} \right\| = \frac{1}{\sqrt{3}}\left\| \begin{matrix} \dot{U}_{BC} \\ \dot{U}_{AB} \\ \dot{U}_{CA} \end{matrix} \right\|. \end{aligned} \tag{4.77}$$

Considering eqn (4.75) and (4.77), the vector representing the load currents can be in the time domain as follows

$$\mathbf{i}(t) = \frac{P}{3U^2}\left\| \begin{matrix} u_{AB}-u_{CA} \\ u_{BC}-u_{AB} \\ u_{CA}-u_{BC} \end{matrix} \right\| + \frac{Q}{\sqrt{3}U^2}\left\| \begin{matrix} u_{BC} \\ u_{CA} \\ u_{AB} \end{matrix} \right\| + \frac{D_R}{3U^2}\left\| \begin{matrix} u_{AB}-u_{CA} \\ u_{CA}-u_{BC} \\ u_{BC}-u_{AB} \end{matrix} \right\|$$
$$+ \frac{D_I}{\sqrt{3}U^2}\left\| \begin{matrix} u_{BC} \\ u_{AB} \\ u_{CA} \end{matrix} \right\|. \tag{4.78}$$

Current decomposition (eqn (4.78)) in the time domain can be used to synthesize a reference signal to the active shunt filter during distributed filtration. Each powers represented by eqn (4.75) and (4.78) can be measured with a power meter. In particular, the effective power settlement equation

$$\begin{aligned} P &= \operatorname{Re}(\bar{\mathbf{i}}^{\wedge}\bar{\mathbf{u}}_P^*) = \operatorname{Re}\left(\dot{I}_A\dot{U}_A^* + \dot{I}_B\dot{U}_B^* + \dot{I}_C\dot{U}_C^*\right) = \\ &= \operatorname{Re}(\dot{I}_A\dot{U}_{AC}^* + \dot{I}_B\dot{U}_{BC}^*) = \int_T (u_{AC}i_A + u_{BC}i_B)dt/T \end{aligned} \tag{4.79}$$

corresponds to the well-established technique of measuring power using two wattmeters connected in the Arcn configuration [59].

The formula of reactive power

$$Q = \operatorname{Re}(\bar{\mathbf{i}}^{\wedge}\bar{\mathbf{u}}_Q^*) = \int_T (u_{EC}i_A + u_{CA}i_B + u_{AB}i_C)dt/T\sqrt{3} \tag{4.80}$$

aligns with the measurement approach using the well-known three wattmeter method [59]. The unbalanced scalar power quantities introduced in this article are calculated in the following way:

$$\begin{aligned} D_R &= \operatorname{Re}(\bar{\mathbf{i}}^{\wedge}\bar{\mathbf{u}}_{DR}^*) = \int_T (u_Ai_A + u_Ci_B + u_Bi_C)dt = \int_T (u_{AB}i_A + u_{CB}i_B)dt/T; \\ D_I &= \operatorname{Re}(\bar{\mathbf{i}}^{\wedge}\bar{\mathbf{u}}_{DI}^*) = \int_T (u_{BC}i_A + u_{AB}i_B + u_{CA}i_C)dt/T\sqrt{3}, \end{aligned} \tag{4.81}$$

which correspond to the measurement setups involving two wattmeters (Figure 4.9a) and three wattmeters (Figure 4.9b), respectively.

With known complex conductances of line load $\bar{Y}_{AB} = G_{AB} + jB_{AB}; \bar{Y}_{BC} = G_{BC} + jB_{BC}; \bar{Y}_{CA} = G_{CA} + jB_{CA}$, when the loads are interconnected in a delta configuration, all the power components can be

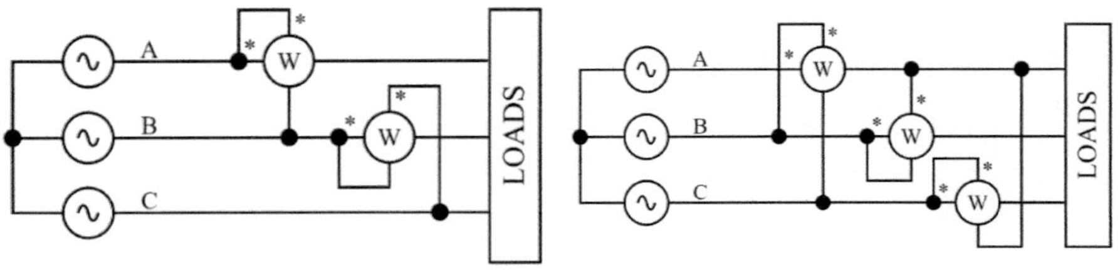

Figure 4.9 (a) Scheme for measuring power D_R.(b) Scheme for measuring power D_I.

expressed through admittance parameters. In fact, under this condition, the following equality holds true:

$$\bar{\mathrm{i}} = \begin{Vmatrix} \dot{U}_{AB}\overline{Y}_{AB} - \dot{U}_{CA}\overline{Y}_{CA} \\ \dot{U}_{BC}\overline{Y}_{BC} - \dot{U}_{AB}\overline{Y}_{AB} \\ \dot{U}_{CA}\overline{Y}_{CA} - \dot{U}_{BC}\overline{Y}_{BC} \end{Vmatrix} = \frac{P - jQ}{U}\bar{\mathbf{e}}_{+} + \frac{D_R - jD_I}{U}\bar{\mathbf{e}}_{-},$$

from which we find expressions for powers

$$P - jQ = U\bar{\mathbf{e}}_{-}^{\wedge}\bar{\mathbf{i}} = U^2(\overline{Y}_{AB} + \overline{Y}_{BC} + \overline{Y}_{CA});$$
$$D_R - jD_I = U\bar{\mathbf{e}}_{+}^{\wedge}\bar{\mathbf{i}} = U^2(e^{j\pi/3}\overline{Y}_{AB} - \overline{Y}_{BC} + e^{-j\pi/3}\overline{Y}_{CA}).$$

Active and reactive powers are determined as follows

$$P = U_L^2(G_{AB} + G_{BC} + G_{CA}); Q = -U_L^2(B_{AB} + B_{BC} + B_{CA}), \qquad (4.82)$$

the expressions coincide with those given in [56]. The new formulas introduced are for determining the values of unbalance power components:

$$D_R = U^2\mathrm{Re}(e^{j\pi/3}\overline{Y}_{AB} - \overline{Y}_{BC} + e^{-j\pi/3}\overline{Y}_{CA});$$
$$D_I = -U^2\mathrm{Im}(e^{j\pi/3}\overline{Y}_{AB} - \overline{Y}_{BC} + e^{-j\pi/3}\overline{Y}_{CA}). \qquad (4.83)$$

The unbalance power module

$$D = \sqrt{D_R^2 + D_I^2} = U^2\left|e^{j\pi/3}\overline{Y}_{AB} - \overline{Y}_{BC} + e^{-j\pi/3}\overline{Y}_{CA}\right|$$
$$= U^2\left|\overline{Y}_{AB} + \dot{a}\overline{Y}_{BC} + \tilde{a}\overline{Y}_{CA}\right|$$

matches the definition in [56] which uses the unbalanced admittance. The benefit of separating the unbalance power into two orthogonal components is the ability to independently control each component during distributed active filtering.

The aim of the virtual experiment is to compare the power values measured by the integral formulas represented by eqn (4.79)–(4.81) with the corresponding estimated values calculated using eqn (4.83). The recent expressions have been modified to account for the complex impedances of the three-phase loads – initially based on the known impedance values $\bar{Z}_{AB} = R_{AB} + jX_{AB}; \bar{Z}_{BC} = R_{BC} + jX_{BC}; \bar{Z}_{CA} = R_{CA} + jX_{CA}$, subsequently, the complex conductances are calculated

$$\overline{Y}_0 = 1/\overline{Z}_{AB} + 1/\overline{Z}_{BC} + 1/\overline{Z}_{CA} = Y_0^R + jY_0^I;$$
$$\overline{Y}_- = e^{j\pi/3}/\overline{Z}_{AB} - 1/\overline{Z}_{BC} + e^{-j\pi/3}/\overline{Z}_{CA} = Y_-^R + jY_-^I, \tag{4.84}$$

and then the following power quantities are determined:

$$P = U^2 Y_0^R; Q = -U^2 Y_0^I; D_R = U^2 Y_-^R; D_I = -U^2 Y_-^I.$$

In the first set of experiments, a linear unbalanced load was connected to a balanced network voltage with parameters $U = 380$ V; $f_{\text{mains}} = 50$ Hz. The load had fixed parameter values in two phases: $R_{BC} = 4\Omega, X_{BC} = -5\Omega \rightarrow C_{BC} = 636.6\mu\text{F}; R_{CA} = 3\Omega, X_{CA} = 3\Omega \rightarrow L_{CA} = 9.54\text{mH}$, while the active resistance of the third phase was varied within the range $0 \leq R_{AB} \leq 10\Omega$.

In the second set of experiments the reactive resistance of this same phase was varied within the range $0 \leq X_{AB} \leq 10\ \Omega$ under constant value of the active resistance ($R_{AB} = 5\ \Omega$). The results of calculating the power components according to eqn (4.84) are presented in the graphs shown in Figure 4.10.

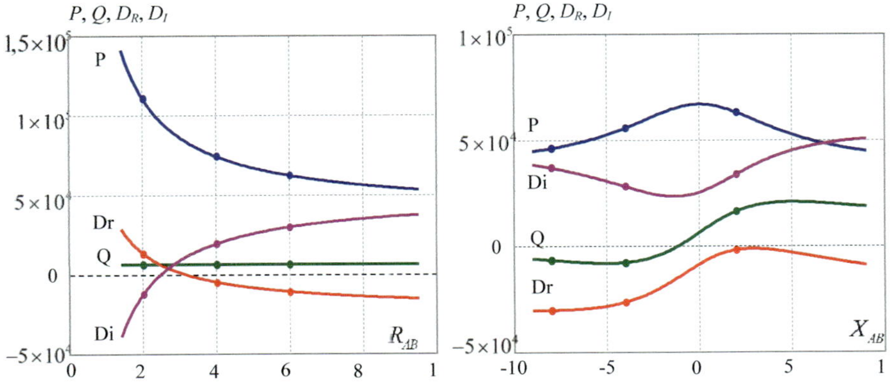

Figure 4.10 Dependencies of powers from R_{AB} when $X_{AB} = 0$, and from X_{AB} when $R_{AB} = 5$.

The data points plotted on the graphs correspond to the power values measured during the simulation experiments. The waveforms of the instantaneous powers and their calculated average values, which were used to generate the data points, are depicted in Figure 4.11.

Figure 4.12 shows the synthesis of the load current waveforms from the power components using eqn (4.78). From this figure, we can observe that the shape of the load current matches the waveform shown in Figure 4.11 during the final period when the average value of the power component has stabilized.

From orthogonal current decomposition (eqn (4.75)) also follows the decomposition of the current root mean square values proportional to the squares of the corresponding powers

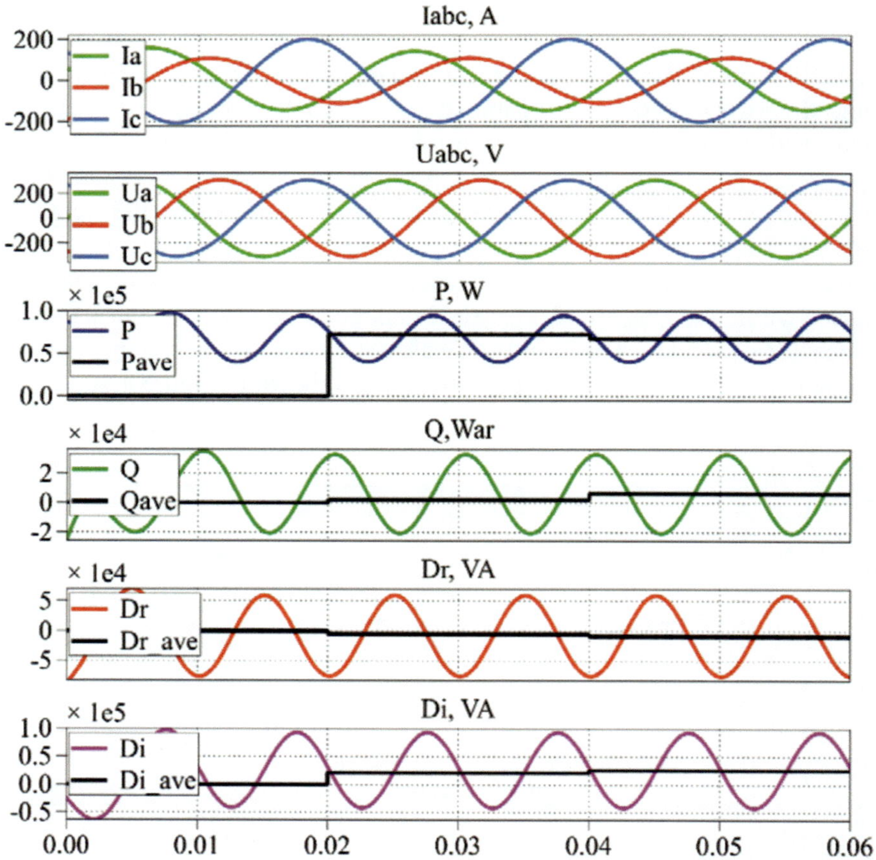

Figure 4.11 The waveforms of instantaneous powers and their average values.

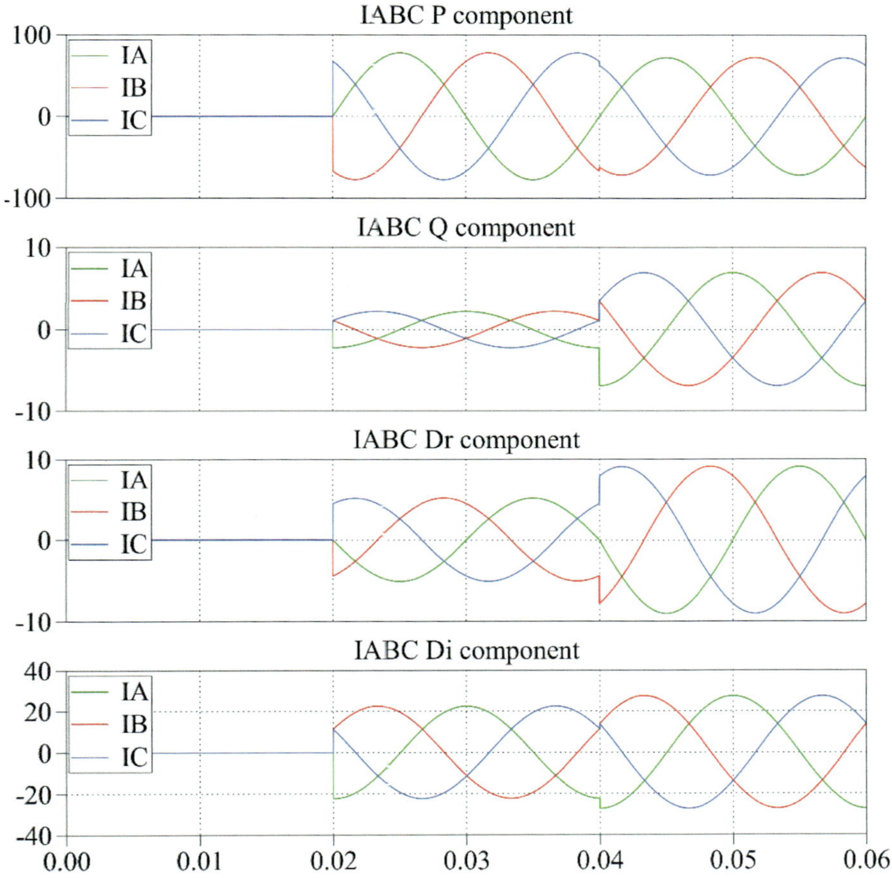

Figure 4.12 The load current waveform based on the calculated power component quantities.

$$I^2 = I_P^2 + I_Q^2 + I_{DR}^2 + I_{DI}^2 = P^2/U^2 + Q^2/U^2 + D_R^2/U^2 + D_I^2/U^2. \quad (4.85)$$

Experimentally confirm that each of the powers given by eqn (4.79)–(4.81) contribute to the current rms values proportional to power losses of the transmission line.

For load $\bar{Z}_{AB} = R_{AB} = 5, \bar{Z}_{BC} = R_{BC} + jX_{BC} = 4 - j, \bar{Z}_{\mathrm{CA}} = R_{\mathrm{CA}} + jX_{CA} = 1 + 4j$,

$\bar{Y}_0 = 1/\bar{Z}_{AB} + 1/\bar{Z}_{BC} + 1/\bar{Z}_{CA} = Y_0^R + jY_0^I = 0,4941 - 0,1765j;$
$\bar{Y}_- = e^{j\pi/3}/\bar{Z}_{AB} - 1/\bar{Z}_{BC} + e^{-j\pi/3}/\bar{Z}_{CA} = Y_-^R + jY_-^I = -0,3097 - 0,0542j,$

values of integrated powers given by eqn (4.79)–(4.81) are

$$P = U^2 Y_0^R = 71687,3; \quad Q = -U^2 Y_0^I = 25602,6;$$
$$D_R = U^2 Y_-^R = -44924,9; \quad D_I = -U^2 Y_-^I = 7864,6,$$

and rms values of current components are

$$I_P^2 = \frac{1}{T}\int_T \mathbf{i}_P^{\wedge}(t)\mathbf{i}_P(t)dt = \bar{\mathbf{I}}_P^{\wedge}\bar{\mathbf{i}}_P^{*} = P^2/U^2 = U^2\left(Y_0^R\right)^2 = 35421,97$$
$$I_Q^2 = Q^2/U^2 = U^2\left(Y_0^I\right)^2 = 4518,11; I_{DR}^2 = D_R^2/U^2 = U^2\left(Y_-^R\right)^2 = 13911,14$$
$$I_{DI}^2 = D_I^2/U^2 = U^2\left(Y_-^I\right)^2 = 426,33; I^2 = I_P^2 + I_Q^2 + I_P^2 + I_Q^2 = 54277,55$$

Instantaneous load currents and power contributions to their rms values are shown in Figure 4.13.

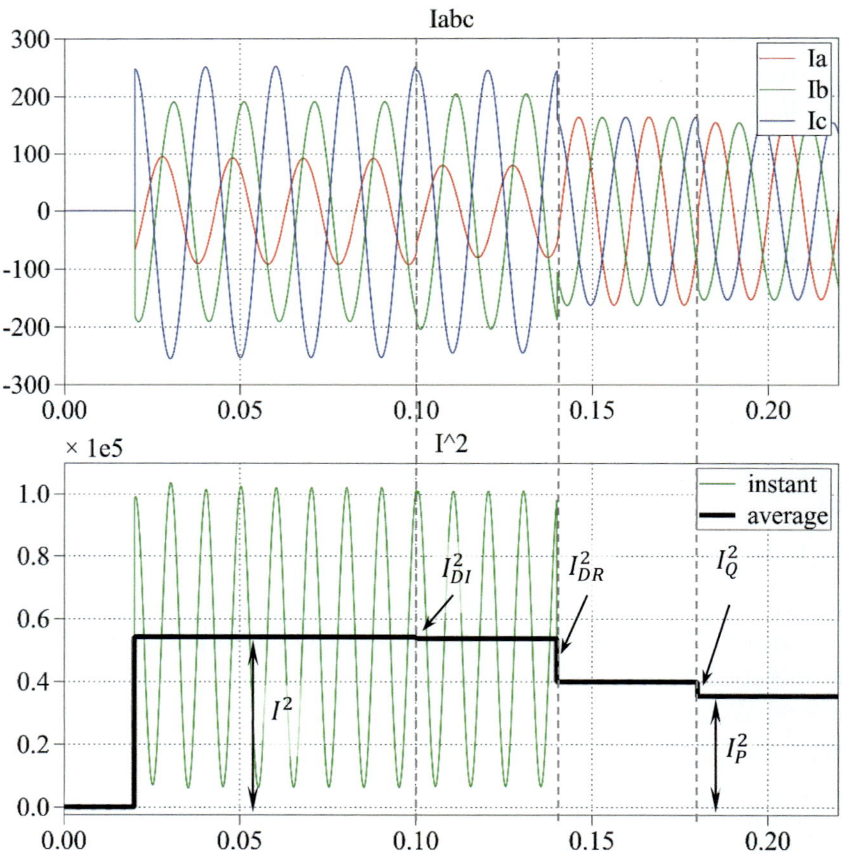

Figure 4.13 Instantaneous load currents and power contributions to their rms values under linear load.

Let us add a nonlinear load to the study circle, preserving the value of line currents according to the first harmonic. To do this, instead of the linear load of phase *AB* $R_{AB} = 5\ \Omega$, we introduce a serial connection of the ideal diode and active resistance $R_{AB}/2 = 2.5\ \Omega$. The instantaneous current of this branch is

$$i_{AB}(t) = u_{AB}(t)/R_{AB} + |u_{AB}(t)|/R_{AB},$$

which ensures the immutability of the first harmonic current parameters and their corresponding powers but causes additional current flowing of higher harmonics in the transmission line

$$\mathbf{i}_H(t) = \begin{Vmatrix} |u_{AB}(t)|/R_{AB} \\ -|u_{AB}(t)|/R_{AB} \\ 0 \end{Vmatrix} = \frac{\sqrt{2}U\,|\cos(\omega t + \pi/6)|}{R_{AB}} \begin{Vmatrix} 1 \\ -1 \\ 0 \end{Vmatrix} \tag{4.86}$$

with an average quadratic value $I_H^2 = \frac{1}{T}\int_T \mathbf{i}_H^{\wedge}(t)\mathbf{i}_H(t)dt = \frac{2U^2}{R_{AB}^2}$.

Let us ensure that this additional component of load currents does not distort the values of integral powers given by eqn (4.79)–(4.81). Figure 4.14 shows graphs of instant powers $p(t)$,$q(t)$,$d_R(t)$,$d_I(t)$, corresponding to the sub-integral expressions of eqn (4.79)–(4.81) and their average values corresponding to the integrated powers P, Q, D_R, D_I for the linear load highlighted in green and for nonlinear load with an additional component of currents (eqn (4.86)) highlighted in blue. We see that instantaneous powers have changed significantly, but there is no difference in their average values. As the following figure (Figure 4.15) shows, their average values are also shown where these components are brought together.

Considering the additional component of line currents corresponding to higher harmonics, the Buchholz apparent power formula takes the form of

$$\mathrm{S}^2 = P^2 + Q^2 + D_R^2 + D_I^2 + H^2,$$

where $H^2 = U^2 I_H^2$ is square of the power of higher harmonics currents.

For the considered load $I_H^2 = \frac{2U^2}{R_{AB}^2} = 11606.52$, and the rms value of nonlinear load currents is equal to

$$I^2 = \frac{1}{T}\int_T \mathbf{i}^{\wedge}(t)\mathbf{i}(t)dt = I_P^2 + I_Q^2 + I_{DR}^2 + I_{DI}^2 + I_H^2 = 65884,07.$$

Let us consider the decomposition of currents and their corresponding powers. Considering the phase voltages as sinusoidal symmetric, we will

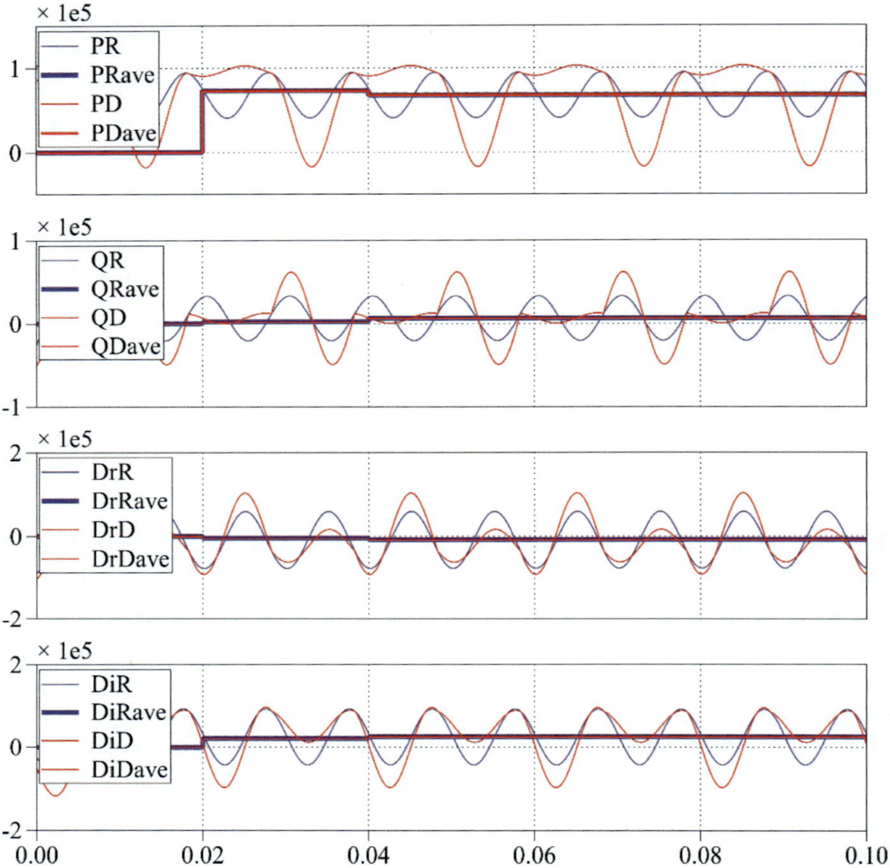

Figure 4.14 Instantaneous power components $p(t)$,$q(t)$,$d_R(t)$,$d_I(t)$ and their average values P, Q, D_R, D_I for linear and nonlinear load.

introduce following complex reference voltage vectors proportional to the unit vectors of symmetric components.

$$\overline{\mathbf{u}}_+ = \overline{\mathbf{u}} = \left\| \begin{matrix} \dot{U}_A \\ \dot{U}_B \\ \dot{U}_C \end{matrix} \right\| = \frac{U}{\sqrt{3}} \left\| \begin{matrix} 1 \\ \tilde{a} \\ \dot{a} \end{matrix} \right\| = U\overline{\mathbf{e}}_+ ; \overline{\mathrm{u}}_- = U\overline{\mathbf{e}}_- = \frac{U}{\sqrt{3}} \left\| \begin{matrix} 1 \\ \dot{a} \\ \tilde{a} \end{matrix} \right\|$$

$$= \left\| \begin{matrix} \dot{U}_A \\ \dot{U}_C \\ \dot{U}_B \end{matrix} \right\| ; \overline{\mathbf{u}}_0 = U\overline{\mathbf{e}}_0 = \frac{U}{\sqrt{3}} \left\| \begin{matrix} 1 \\ 1 \\ 1 \end{matrix} \right\| = \left\| \begin{matrix} \dot{U}_A \\ \dot{U}_A \\ \dot{U}_A \end{matrix} \right\| .$$

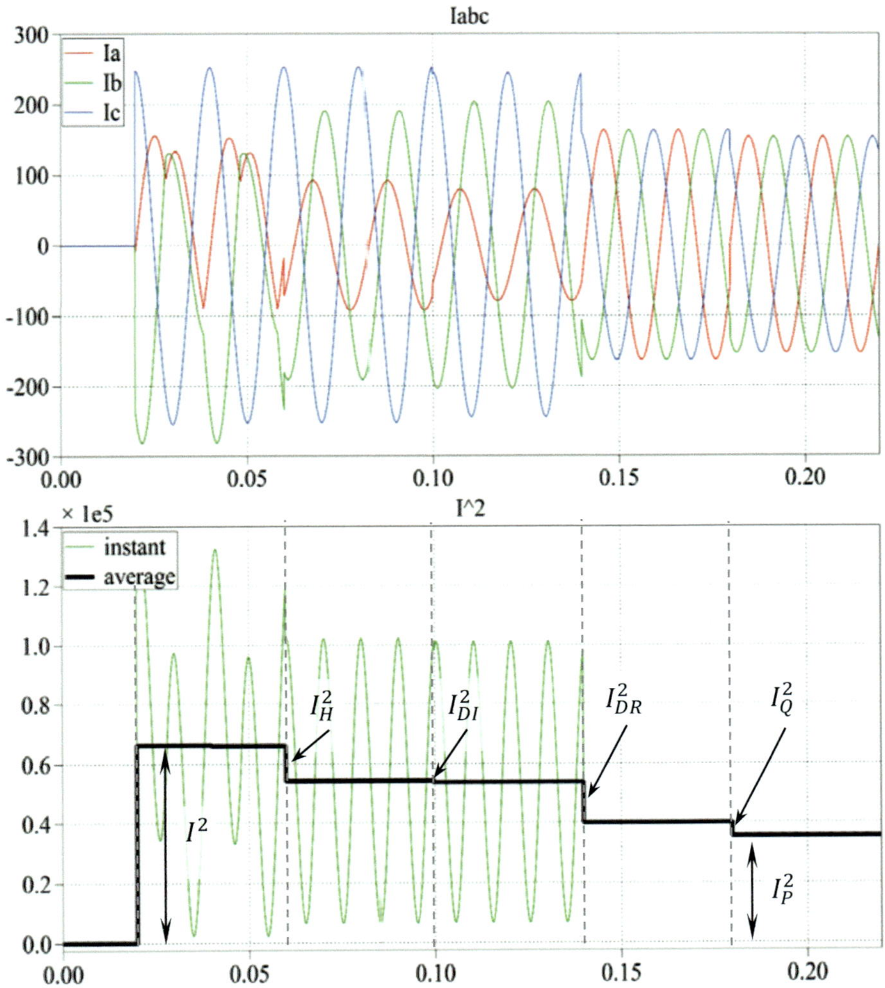

Figure 4.15 Instantaneous load currents and power contributions to their rms values under nonlinear load.

Then the three-coordinate current vector from linear load allows such decomposition into symmetric components in the frequency domain

$$\bar{\mathbf{i}} = \left\| \begin{matrix} \dot{I}_A \\ \dot{I}_B \\ \dot{I}_C \end{matrix} \right\| = \left\| \begin{matrix} I_A e^{j\psi_A} \\ I_B e^{j\psi_B} \\ I_C e^{j\psi_C} \end{matrix} \right\| = \frac{\dot{I}_+}{\sqrt{3}} \left\| \begin{matrix} 1 \\ \tilde{a} \\ \dot{a} \end{matrix} \right\| + \frac{\dot{I}_-}{\sqrt{3}} \left\| \begin{matrix} 1 \\ \dot{a} \\ \tilde{a} \end{matrix} \right\| + \frac{\dot{I}_0}{\sqrt{3}} \left\| \begin{matrix} 1 \\ 1 \\ 1 \end{matrix} \right\|$$
$$= (\dot{I}_+ \bar{\mathbf{u}}_+ + \dot{I}_- \bar{\mathbf{u}}_- + \dot{I}_0 \bar{\mathbf{u}}_0) U^{-1}. \quad (4.87)$$

Complex scalar coefficients of decomposition (eqn (4.87)) will be found as scalar products of the load current vector and corresponding complex reference voltage vectors

$$\begin{aligned}
\bar{\mathbf{i}}^{\wedge}\bar{\mathbf{u}}_{+}^{*} &= P - jQ = \dot{I}_{+}U; I_{R+} = P/U; I_{I+} = -Q/U;\\
\bar{\mathbf{i}}^{\wedge}\bar{\mathbf{u}}_{-}^{*} &= D_R - jD_I = \dot{I}_{-}U; I_{R-} = D_R/U; I_{I-} = -D_I/U; \\
\bar{\mathbf{i}}^{\wedge}\bar{\mathbf{u}}_{0}^{*} &= N_R - jN_I = \dot{I}_{0}U; I_{R0} = \frac{N_R}{U}; I_{I0} = -\frac{N_I}{U}.
\end{aligned} \quad (4.88)$$

Then the current decomposition is

$$\begin{aligned}
\bar{\mathbf{i}} &= \begin{Vmatrix}\dot{I}_A\\ \dot{I}_B\\ \dot{I}_C\end{Vmatrix} = \frac{P-jQ}{U_\Phi U\sqrt{3}}\begin{Vmatrix}\dot{U}_A\\ \dot{U}_B\\ \dot{U}_C\end{Vmatrix} + \frac{D_R-jD_I}{U_\Phi U\sqrt{3}}\begin{Vmatrix}\dot{U}_A\\ \dot{U}_C\\ \dot{U}_B\end{Vmatrix} + \frac{N_R-jN_I}{U_\Phi U\sqrt{3}}\begin{Vmatrix}\dot{U}_A\\ \dot{U}_A\\ \dot{U}_A\end{Vmatrix} =\\
&= \frac{P}{U^2}\begin{Vmatrix}\dot{U}_A\\ \dot{U}_B\\ \dot{U}_C\end{Vmatrix} - \frac{Q}{U^2}\begin{Vmatrix}j\dot{U}_A\\ j\dot{U}_B\\ j\dot{U}_C\end{Vmatrix} + \frac{D_R}{U^2}\begin{Vmatrix}\dot{U}_A\\ \dot{U}_C\\ \dot{U}_B\end{Vmatrix} - \frac{D_I}{U^2}\begin{Vmatrix}j\dot{U}_A\\ j\dot{U}_C\\ j\dot{U}_B\end{Vmatrix} + \frac{N_R}{U^2}\begin{Vmatrix}\dot{U}_A\\ \dot{U}_A\\ \dot{U}_A\end{Vmatrix}\\
&- \frac{N_I}{U^2}\begin{Vmatrix}j\dot{U}_A\\ j\dot{U}_A\\ j\dot{U}_A\end{Vmatrix} = \frac{P}{U^2}\begin{Vmatrix}\dot{U}_A\\ \dot{U}_B\\ \dot{U}_C\end{Vmatrix} + \frac{Q}{U^2\sqrt{3}}\begin{Vmatrix}\dot{U}_{BC}\\ \dot{U}_{CA}\\ \dot{U}_{AB}\end{Vmatrix} + \frac{D_R}{U^2}\begin{Vmatrix}\dot{U}_A\\ \dot{U}_C\\ \dot{U}_B\end{Vmatrix} +\\
&+ \frac{D_I}{U^2\sqrt{3}}\begin{Vmatrix}\dot{U}_{BC}\\ \dot{U}_{AB}\\ \dot{U}_{CA}\end{Vmatrix} + \frac{N_R}{U^2}\begin{Vmatrix}\dot{U}_A\\ \dot{U}_A\\ \dot{U}_A\end{Vmatrix} + \frac{N_I}{U^2\sqrt{3}}\begin{Vmatrix}\dot{U}_{BC}\\ \dot{U}_{BC}\\ \dot{U}_{BC}\end{Vmatrix} =\\
&= (P\bar{\mathbf{u}}_P + Q\bar{\mathbf{u}}_Q + D_R\bar{\mathbf{u}}_{DR} + D_I\bar{\mathbf{u}}_{DI} + N_R\bar{\mathbf{u}}_{NR} + N_I\bar{\mathbf{u}}_{\mathbf{N}I})\,U^{-2} =\\
&= \bar{\mathbf{i}}_P + \bar{\mathbf{i}}_Q + \bar{\mathbf{i}}_{DR} + \bar{\mathbf{i}}_{DI} + \bar{\mathbf{i}}_{NR} + \bar{\mathbf{i}}_{NI}.
\end{aligned}$$

Since all components of currents are mutually orthogonal, power extraction can be carried out by wattmeters, and expressions for integral powers P, Q, D_R, D_I correspond to eqn (4.79)–(4.81), and one wattmeter is enough to determine each of the integral powers N_R, N_I, caused by nonzero neutral current:

$$\begin{aligned}
N_I &= \mathrm{Re}(\bar{\mathbf{i}}^{\wedge}\bar{\mathbf{u}}_{NI}^{*}) = \int_T (u_A i_A + u_A i_B + u_A i_C)dt/T =\\
&= \int_T u_A(i_A + i_B + i_C)dt/T = \int_T u_A i_N dt/T;\\
N_R &= \mathrm{Re}(\bar{\mathbf{i}}^{\wedge}\bar{\mathbf{u}}_{NR}^{*}) = \int_T (u_{BC} i_A + u_{BC} i_B + u_{BC} i_C)dt/T\sqrt{3}
\end{aligned}$$

$$= \int_T u_{BC} i_N dt / T\sqrt{3}. \tag{4.89}$$

Decomposition of the first harmonic current vector in the time domain takes the form

$$\begin{aligned}
\mathbf{i}_1(t) &= \begin{Vmatrix} i_{1A} \\ i_{1B} \\ i_{1C} \end{Vmatrix} = \frac{P}{U^2} \begin{Vmatrix} u_A \\ u_B \\ u_C \end{Vmatrix} \\
&+ \frac{Q}{U^2\sqrt{3}} \begin{Vmatrix} u_{BC} \\ u_{CA} \\ u_{AB} \end{Vmatrix} + \frac{D_R}{U^2} \begin{Vmatrix} u_A \\ u_C \\ u_B \end{Vmatrix} + \qquad (4.90) \\
&+ \frac{D_I}{U^2\sqrt{3}} \begin{Vmatrix} u_{BC} \\ u_{AB} \\ u_{CA} \end{Vmatrix} + \frac{N_R}{U^2} \begin{Vmatrix} u_A \\ u_A \\ u_A \end{Vmatrix} + \frac{N_I}{U^2\sqrt{3}} \begin{Vmatrix} u_{BC} \\ u_{BC} \\ u_{BC} \end{Vmatrix} = \\
&= (P\mathbf{u}_P + Q\mathbf{u}_Q + D_R\mathbf{u}_{DR} + D_I\mathbf{u}_{DI} + N_R\mathbf{u}_{NR} + N_I\mathbf{u}_{NI})U^{-2} = \\
&= \mathbf{i}_P + \mathbf{i}_Q + \mathbf{i}_{DR} + \mathbf{i}_{DI} + \mathbf{i}_{NR} + \mathbf{i}_{NI}.
\end{aligned}$$

According to this expression, the compensation signal for the main harmonics during shunt distributed filtration can be synthesized which contains current terms of inactive powers.

In the presence of current harmonics in the four-wire transmission lines, the load current vector has the form of $\mathbf{i}(t) = \mathbf{i}_1(t) + \mathbf{i}_H(t)$, the eqn (4.79)–(4.81), (4.89) for inactive powers do not change, and the power loss is

$$\begin{aligned}
\Delta P &= \frac{1}{T}\int_T [i_A^2(t)r + i_B^2(t)r + i_C^2(t)r + i_N^2(t)r_N]dt = \\
&= \frac{1}{T}\int_T \| \; i_A \quad i_B \quad i_C \; \| \begin{Vmatrix} r+r_N & r_N & r_N \\ r_N & r+r_N & r_N \\ r_N & r_N & r+r_N \end{Vmatrix} \begin{Vmatrix} i_A \\ i_B \\ i_C \end{Vmatrix} dt = \\
&= \frac{r}{T}\int_T \mathbf{i}^{\wedge}(t)\left(\mathbf{I} + 3\rho\bar{\mathbf{e}}_0\bar{\mathbf{e}}_0^{\wedge}\right)\mathbf{i}(t)\,dt = r\bar{\mathbf{i}}^{\wedge}\left(\mathbf{I} + 3\rho\bar{\mathbf{e}}_0\bar{\mathbf{e}}_0^{\wedge}\right)\bar{\mathbf{i}}^{*} + \Delta P_H = \\
&= r[\bar{\mathbf{i}}^{\wedge}\bar{\mathbf{i}}^{*} + 3\rho(\bar{\mathbf{i}}^{\wedge}\bar{\mathbf{e}}_0)(\bar{\mathbf{i}}^{\wedge}\bar{\mathbf{e}}_0)^{*}] + \Delta P_H = \\
&= r\left[\left(P^2 + Q^2 + D_R^2 + D_I^2 + N_R^2 + N_I^2\right)U^{-2} + 3\rho \times \frac{(N_R - jN_I)\,U_\Phi\sqrt{3}}{U_\Phi U\sqrt{3}} \times \right. \\
&\quad \left. \times \frac{(N_R - jN_I)^*}{U}\right] + \Delta P_H = \\
&= [P^2 + Q^2 + D_R^2 + D_I^2 + (N_R^2 + N_I^2)(1 + 3\rho) + H^2]rU^{-2},
\end{aligned}$$

where $\rho = r_N/r$ is the resistive parameter ratio of the transmission line; ΔP_H- power losses of higher harmonics; $H^2 = U^2\Delta P_H/r$.

Short-circuit power is equal to

$$P_0 = \frac{1}{T}\int_T [u_A^2(t)/r + u_B^2(t)/r + u_C^2(t)/r]dt = 3U_\Phi^2/r = U^2/r.$$

The square of apparent power and its decomposition on quadratic components

$$S^2 = \Delta PP_0 = P^2 + Q^2 + D_R^2 + D_I^2 + (N_R^2 + N_I^2)(1 + 3\rho) + H^2 \qquad (4.91)$$

differs from known formulas of other power theories by the presence of a multiplier $(1 + 3\rho)$. This is caused by the negative influence of current components corresponding to the powers N_R, N_I which create additional power losses in neutral wire of three-phase four-wire power system.

For the possibility of experimental verification of the current decomposition (eqn (4.90)) and the power decomposition (eqn (4.91)) we first consider linear load connected to the star that is described by complex conductivities $\overline{Y}_A, \overline{Y}_B$, and $\overline{Y}_C$. The current vector caused by this load is

$$\bar{\mathbf{i}} = \begin{Vmatrix} \dot{I}_A \\ \dot{I}_B \\ \dot{I}_C \end{Vmatrix} = \begin{Vmatrix} \dot{U}_A\overline{Y}_A \\ \dot{U}_B\overline{Y}_B \\ \dot{U}_C\overline{Y}_C \end{Vmatrix} = U_\Phi \begin{Vmatrix} \overline{Y}_A \\ \tilde{a}\overline{Y}_B \\ \dot{a}\overline{Y}_C \end{Vmatrix}.$$

Complex powers according to eqn (4.88)

$$P - jQ = \bar{\mathbf{i}}^\wedge \overline{\mathbf{u}}_+^* = U_\Phi \begin{Vmatrix} \overline{Y}_A \\ \tilde{a}\overline{Y}_B \\ \dot{a}\overline{Y}_C \end{Vmatrix}^\wedge \times U_\Phi \begin{Vmatrix} 1 \\ \tilde{a} \\ \dot{a} \end{Vmatrix}^* = U_\Phi^2(\overline{Y}_A + \overline{Y}_B + \overline{Y}_B);$$

$$D_R - jD_I = \bar{\mathbf{i}}^\wedge \overline{\mathbf{u}}_-^* = U_\Phi \begin{Vmatrix} \overline{Y}_A \\ \tilde{a}\overline{Y}_B \\ \dot{a}\overline{Y}_C \end{Vmatrix}^\wedge \times U_\Phi \begin{Vmatrix} 1 \\ \dot{a} \\ \tilde{a} \end{Vmatrix}^* = U_\Phi^2(\overline{Y}_A + \dot{a}\overline{Y}_B + \tilde{a}\overline{Y}_B);$$

$$N_R - jN_I = \bar{\mathbf{i}}^\wedge \overline{\mathbf{u}}_0^* = U_\Phi \begin{Vmatrix} \overline{Y}_A \\ \tilde{a}\overline{Y}_B \\ \dot{a}\overline{Y}_C \end{Vmatrix}^\wedge \times U_\Phi \begin{Vmatrix} 1 \\ 1 \\ 1 \end{Vmatrix}^* = U_\Phi^2(\overline{Y}_A + \tilde{a}\overline{Y}_B + \dot{a}\overline{Y}_B),$$

have the following numerical values for

$$\overline{Z}_A = R_A = 5, \overline{Z}_B = R_B + jX_B = 4 - j, \overline{Z}_C = R_C + jX_C = 1 + 4j:$$

$$P = U_\Phi^2 \mathrm{Re}\left(\overline{Y}_A + \overline{Y}_B + \overline{Y}_C\right) = 23895,78;$$
$$Q = -U_\Phi^2 \mathrm{Im}\left(\overline{Y}_A + \overline{Y}_B + \overline{Y}_C\right) = 8534,21;$$
$$D_R = U_\Phi^2 \mathrm{Re}\left(\overline{Y}_A + \dot{a}\overline{Y}_B + \tilde{a}\overline{Y}_C\right) = -9757,8;$$
$$D_I = -U_\Phi^2 \mathrm{Im}(\overline{Y}_A + \dot{a}\overline{Y}_B + \tilde{a}\overline{Y}_C) = -11657,94;$$
$$N_R = U_\Phi^2 \mathrm{Re}\left(\overline{Y}_A + \tilde{a}\overline{Y}_B + \dot{a}\overline{Y}_C\right) = 14878,33;$$
$$N_I = -U_\Phi^2 \mathrm{Im}(\overline{Y}_A + \tilde{a}\overline{Y}_B + \dot{a}\overline{Y}_C) = 3123,74.$$

Modeling of integral expressions

$$\begin{aligned}
P &= \frac{1}{T}\int_T (u_A i_A + u_B i_B + u_C i_C)dt; Q = \\
&= \frac{1}{\sqrt{3}T}\int_T (u_{BC} i_A + u_{CA} i_B + u_{AB} i_C)dt; N_R = \frac{1}{T}\int_T u_A i_N dt; \\
D_R &= \frac{1}{T}\int_T (u_A i_A + u_C i_B + u_B i_C)dt; D_I = \\
&= \frac{1}{\sqrt{3}T}\int_T (u_{BC} i_A + u_{AB} i_B + u_{CA} i_C)dt; N_I = \frac{1}{\sqrt{3}T}\int_T u_{BC} i_N dt
\end{aligned}$$

gives similar numerical values.

The active current, which provides the required active loading power, determines the minimum possible relative power losses

$$\Delta P_{\mathrm{MIN}}/r = \frac{1}{T}\int_T \mathbf{i}_P^\wedge(t)\mathbf{i}_P(t)dt = I_P^2 = P^2U^{-2} = 3935,77.$$

Each of the inactive components of linear decomposition currents (eqn (4.90)) determines its additive contribution to the total power losses

$$\begin{aligned}
I_Q^2 &= \Delta P_Q/r = \frac{1}{T}\int_T \mathbf{i}_Q^\wedge(t)\mathbf{i}_Q(t)dt = Q^2U^{-2} = 502.01; \\
I_{DR}^2 &= \Delta P_{DR}/r = \frac{1}{T}\int_T \mathbf{i}_{DR}^\wedge(t)\mathbf{i}_{DR}(t)dt = D_R^2U^{-2} = 656.28; \\
I_{DI}^2 &= \Delta P_{DI}/r = \frac{1}{T}\int_T \mathbf{i}_{DI}^\wedge(t)\mathbf{i}_{DI}(t)dt = D_I^2U^{-2} = 936.77; \\
I_{NR}^2 &= \Delta P_{NR}/r = \frac{1}{T}\int_T \mathbf{i}_{NR}^\wedge(t)(\mathbf{I} + 3\rho\overline{e}_0\overline{e}_0^\wedge)\mathbf{i}_{NR}(t)dt = (1+3\rho)N_R^2U^{-2} \\
&= 3814.49;
\end{aligned}$$

$$I_{NI}^2 = \Delta P_{NI}/r = \frac{1}{T}\int_T \mathbf{i}_{NI}^{\wedge}(t)(\mathbf{I}+3\rho\bar{\mathrm{e}}_0\bar{\mathrm{e}}_0^{\wedge})\mathbf{i}_{NI}(t)dt = (1+3\rho)N_I^2U^{-2}$$
$$= 168.14.$$

Replacing of the phase A linear load by a branch containing a series connected active resistance $R_A/2 = 2.5\ \Omega$ and the ideal diode causes the flow in the transmission line of additional currents of higher harmonics, described by the instantaneous values vector $\mathbf{i}_H(t) = |u_A(t)|/R_{AB}\| \ 1 \quad 0 \quad 0 \ \|^{\wedge}$. It causes the additional relative power losses

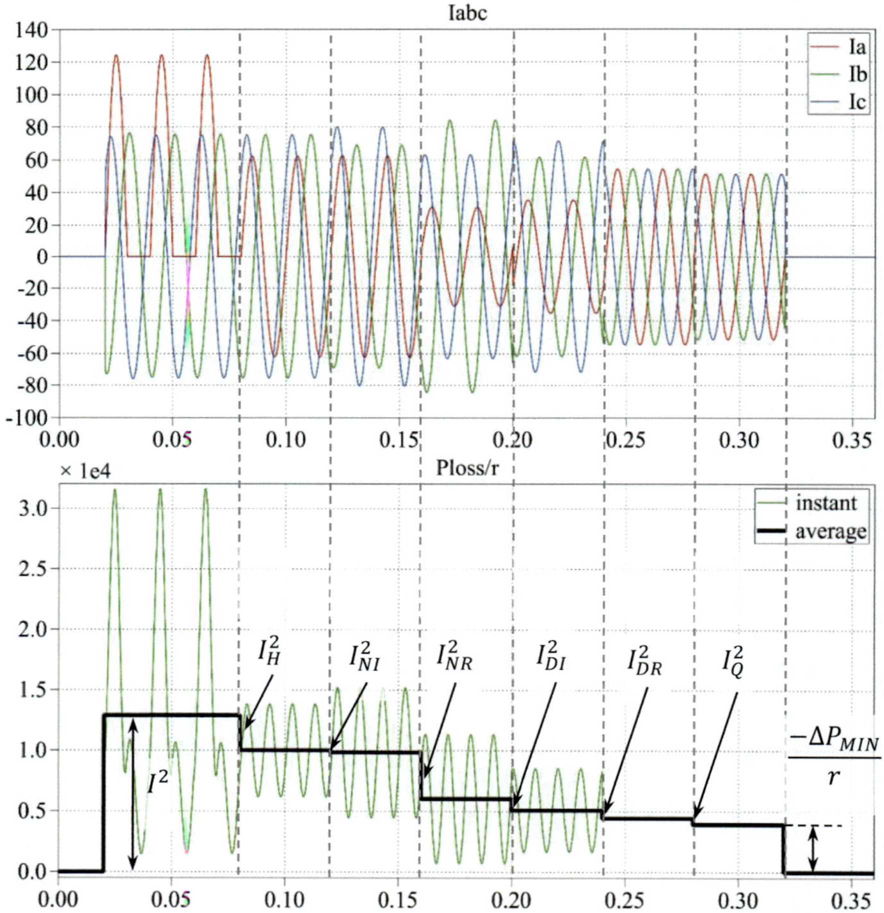

Figure 4.16 Instantaneous load currents and power contributions to their rms values in three-phase four-wire power system.

$$\Delta P_H/r = I_H^2 = \frac{r+r_N}{rT}\int_T \mathbf{i}_H^{\wedge}(t)\mathbf{i}_H(t)dt = \frac{U_\Phi^2(1+\rho)}{R_A^2} = 2901,63.$$

According to eqn (4.91) the relative total power losses will be

$$\Delta P/r = I_P^2+I_Q^2+I_{DR}^2+I_{DI}^2+I_{NR}^2+I_{NI}^2+I_H^2 = I^2 = 12915,1.$$

Figure 4.16 displays the time-varying waveforms of the instantaneous current and power values, along with the relative power losses associated with each quadratic component that makes up the total apparent power, for the case of a nonlinear load.

4.5 Conclusions

1. The instantaneous power theory for multiphase power systems was further extended in [52, 53, 55, 56] by deriving new formulas for instantaneous active current, instantaneous apparent power, and instantaneous power factor. These new formulas accounted for the dependence on the ratio of resistances among the conductors in the transmission line. If the zero-sequence components of the current and voltage vectors are zero, or if the neutral wire resistance is zero, then the expressions reduce to the classical formulas of instantaneous power theory.
2. The strategy of active filtering that forming an instantaneous active current (eqn (4.37)) in the transmission line of multiphase power supply system was proposed. In the presence of instantaneous zero-sequence voltage component, this strategy prevails in the instantaneous power losses strategies based on the orthogonal decomposition of load currents using *pq*-theory of instantaneous power and its modification, cross-vector theory, *pqr*-theory of instantaneous power, proportional-vector theory with deprivation of the zero-sequence voltage component.
3. The new formula (eqn (4.50)) for the apparent power of a multiphase power system using a transmission line resistance model is proposed. This is the geometric mean of the allowable power loss in the transmission line and the short-circuit power of the multiphase power supply on the transmission line resistances.
4. The formation of the active current vector represented by eqn (4.37) by shunt active filter in the transmission line of a multiphase power supply system provides a minimum power loss and a unit value of the power factor. This filtering strategy outperforms other known filtering

strategies based on active current formulas authored by S. Fryze and P. Salmeron / J. Montano.

5. Comparative analysis showed that the proposed formula of apparent power in three-phase four-wire power system is equivalent to that given by the IEEE standard [39] when performing a ratio (eqn (4.68)), which eliminates the uncertainty of the choice of coefficient ξ in the specified standard, and fully corresponds to the European approach to determining apparent power for the case of sinusoidal power supply mode.
6. Using the proposed integrated formula of apparent power, calculated ratios have been found to minimize power loss when using energy-efficient control algorithms for shunt active filtration that correspond to apparent power decompositions regulated by the IEEE standard [39].
7. Two components are proposed to identify the unbalance power of three-phase three-wire power system, each of which has orthogonal current component and determines a separate quadratic contribution to the squared apparent power. These can be used for separate compensation in active distributed filtration.
8. For a given symmetrical voltage source, the sinusoidal process of a three-phase, four-wire power system with asymmetrical loads is found to be completely specified by six scalar values (active and reactive power, the orthogonal components of unbalance power, and the orthogonal component of power associated with the neutral line current). These scalar powers can be used to synthesize control signals for active filtration in grid-connected distributed generation systems.
9. Integral formulas for determining the inactive power were obtained, allowing the contribution of electrical energy to quality degradation to be recorded. These equations were shown to maintain the output value of the reactive power even in the presence of current harmonics. The relationship between the unbalance power component and linear unbalanced load parameters was established, and computer experiments allowed validation of these integral equations.
10. The squares of inactive powers associated with neutral wire current include apparent power decomposition and power loss with multiplier, which depends on the ratio of the resistance transmission line resistances.

References

[1] F. Blaabjerg, Z. Chen and S.B. Kjaer, 'Power electronics as efficient interface in dispersed power generation systems', IEEE Transaction on Power Electronics, vol. 19, no. 5, pp. 1184–1194, Sep. 2004.

[2] Kranthi Kumar Vanukuru, B. Pakkiraiah, 'Distributed Generation Integration to Grid using DG Controlled PQ Theory', International Journal of Engineering and Advanced Technology, vol. 8, iss. 6S3, Sept. 2019.

[3] A. Egea-Alvarez, A. Junyent-Ferré, O. Gomis-Bellmunt, 'Active and Reactive Power Control of Grid Connected Distributed Generation Systems', In: Wang L. (eds) Modeling and Control of Sustainable Power Systems. Green Energy and Technology. Springer, Berlin, Heidelbergb 2012.

[4] F. Harirchi and M. G. Simões, 'Enhanced Instantaneous Power Theory Decomposition for Power Quality Smart Converter Applications', in IEEE Transactions on Power Electronics, vol. 33, no. 11, pp. 9344-9359, Nov. 2018.

[5] Augusto Matheus dos Santos Alonso, Danilo Iglesias Brandao, Elisabetta Tedeschi, Fernando Pinhabel Marafão, 'Distributed selective harmonic mitigation and decoupled unbalance compensation by coordinated inverters in three-phase four-wire low-voltage networks', Electric Power Systems Research, Vol. 186, 2020, 106407.

[6] D. Kumar and F. Zare, 'Harmonic analysis of grid connected power electronic systems in low voltage distribution networks', IEEE J. Emerg. Sel. Topics Power Electron., vol. 4, no. 1, pp. 70–79, Mar. 2016.

[7] F. Harirchi and M. G. Simões, 'Enhanced Instantaneous Power Theory Decomposition for Power Quality Smart Converter Applications', in IEEE Transactions on Power Electronics, vol. 33, no. 11, pp. 9344–9359, Nov. 2018.

[8] P.A. Karafotis, V.A. Evangelopoulos, P.S. Georgilakis, 'Evaluation of harmonic contribution to unbalance in power systems under non-stationary conditions using wavelet packet transform', Electr. Power Syst. Res. 178 (2020) 106026.

[9] S.Y.M. Mousavi, A. Jalilian, M. Savaghebi, J.M. Guerrero, 'Power quality enhancement and power management of a multifunctional interfacing inverter for PV and battery energy storage system', Int. Trans. Electr. Energy Syst. vol. 28, iss. 12, 2018.

[10] Z. Cheng, J. Duan and M.-Y. Chow, 'To Centralize or to Distribute: That Is the Question: A Comparison of Advanced Microgrid Management Systems', in IEEE Industrial Electronics Magazine, vol. 12, no. 1, pp. 6–24, March 2018.

[11] J.M. Guerrero, M. Chandorkar, T. Lee and P.C. Loh, 'Advanced Control Architectures for Intelligent Microgrids – Part I: Decentralized and Hierarchical Control', in IEEE Transactions on Industrial Electronics, vol. 60, no. 4, pp. 1254–1262, April 2013.

[12] V. Monteiro, J.L. Afonso, 'A Unified Topology for the Integration of Electric Vehicle, Renewable Energy Source, and Active Filtering for the Power Quality Improvement of the Electrical Power Grid: An Experimental Validation', Electronics, vol. 11, 429, 2022.

[13] B. Wang, P. Dehghanian and D. Zhao, 'Chance-Constrained Energy Management System for Power Grids with High Proliferation of Renewables and Electric Vehicles', in IEEE Transactions on Smart Grid, vol. 11, no. 3, pp. 2324–2336, May 2020.

[14] W. Kempton, J. Tomic, 'Vehicle-to-Grid Power Implementation: From Stabilizing the Grid to Supporting Large-Scale Renewable Energy', Elsevier J. Power Sources, vol. 144, pp. 280–294, 2015.

[15] J.A. Peças Lopes, F.J. Soares, P.M. Almeida, M.M. da Silva, 'Smart Charging Strategies for Electric Vehicles: Enhancing Grid Performance and Maximizing the Use of Variable Renewable Energy Resources', In Proceedings of the 24th International Battery, Hybrid and Fuel Cell Electric Vehicle Symposium & Exhibition, Stavanger, Norway, 13–16 May 2009; vol. 4, pp. 2680–2690.

[16] S. Gao, K. T. Chau, C. Liu, D. Wu and C.C. Chan, 'Integrated Energy Management of Plug-in Electric Vehicles in Power Grid with Renewables', in IEEE Transactions on Vehicular Technology, vol. 63, no. 7, pp. 3019-3027, Sept. 2014.

[17] Emanuel A. E. Power definitions and the physical mechanism of power flow. John Wiley & Sons. IEEE Press. 2010. 274 p.

[18] M.Yu. Artemenko, Y.V. Kutafin, V.M. Mykhalskyi, S.Y. Polishchuk, V.V. Chopyk, I.A. Shapoval, 'Theory of instantaneous power supply systems taking into account resistive parameters of the transmission line', Technical electrodynamics, no. 4, pp. 12–22, 2019.

[19] R.S. Herrera, P. Salmeron, 'Instantaneous reactive power theory: A Comparative Evaluation of different formulations', IEEE Trans. Power Delivery, vol. 22, no. 1, pp. 595–604, 2007.

[20] M.Yu. Artemenko, L.M. Batrak, 'Multivariate proportional-vector control system of parallel active filter of four-wire three-phase power supply system', Technical electrodynamics, no. 5, pp. 83–85, 2014.

[21] M.Yu. Artemenko, L.M. Batrak, V.M. Mykhalskyi, S.Y. Polishchuk, 'Analysis of the possibility of increasing the efficiency of a three-phase four-wire power supply system by means of parallel active filtration', Technical electrodynamics, no. 6, pp. 12–18, 2015.

[22] J.G. Mayordomo, J. Usaola, 'Apparent power and power factor definitions for polyphase non-linear loads when supply conductors present different resistances', European Transactions on Electrical Power, vol. 3, no. 6, pp. 415–420, Nov/Dec 1993.

[23] P. Salmeron, R.S. Herrera, 'Instantaneous Reactive Power Theory – A General Approach to Poly-Phase Systems', Electric Power Systems Research, vol.79(2009), pp. 1263–1270, 2009.

[24] M.Yu. Artemenko, V.M. Mykhalskyi, S.Y. Polishchuk, V.V. Chopyk, I.A. Shapoval, 'The Instantaneous Power Theory of Multiphase Power Supply Systems and Its Application to Energy-Saving Shunt Active Filtering', Proceedings of the IEEE 2nd Ukraine Conference on Electrical and Computer Engineering (UKRCON) 2019. Lviv (Ukraine), pp. 334–340, July 2–6 2019.

[25] G. Korn, T. Korn, Casekeeper for mathematics for researchers and engineers. M: Science. 1974. 822 p.

[26] S.Y. Polishchuk, M.Yu. Artemenko, V.M. Mykhalskyi, I.A. Shapoval, L.M. Batrak, 'Strategy of control of parallel active fi-ltre with partial weakening of the component of zero sequence on the edge of the three-phase four-wire network', Technical electrodynamics, no. 3, pp. 12–19, 2013.

[27] A. Garcesa, M. Molinas, P. Rodriguez, 'A generalized compensa-tion theory for active filters based on mathematical optimization in ABC frame', Electric Power Systems Research, vol. 90, pp. 1–10, 2012.

[28] F.Z. Peng, J.S. Lai, 'Generalized instantaneous reactive power theory of three-phase power systems', IEEE Trans. Instrum. Meas, vol. 45, no. 1, pp. 293–297, 1996.

[29] Yu.A. Sirotin, 'Vectorial instantaneous power and energy modes in three-phase circuits', Technical electrodynamics, no. 6, pp. 57–65, 2013.

[30] M.Yu. Artemenko, L.M. Batrak, 'Method of calculating the parameters of the reactive compensator of the parallel hybrid filter for a random discrete linear load model'. Bulletin of the Kyiv National University of Technology and Design, no. 6, pp. 115–121, 2015.

[31] M.Yu. Artemenko, 'Apparent power of three-phase power supply system in non-sinusoidal mode and energy efficiency of parallel active filtration'. Electronics and communications, no. 6, pp. 38–47, 2014.
[32] M.Yu. Artemenko, V.M. Mykhalskyi, S.Y. Polishchuk, 'Definition of apparent power of three-phase power supply systems as a theoretical basis for the construction of energy-efficient means of parallel active filtration', no. 2, pp. 25–34, 2017.
[33] M. Artemenko, L. Batrak, S. Polishchuk, 'New definition formulas for apparent power and active current of three-phase power system [Nowa definicja mocy pozornej i prądu czynnego w układzie trójfazowym]', Przeglad Elektrotechniczny, no. 95(8), pp. 81–85, 2019.
[34] M.Yu. Artemenko, L.M. Batrak, S.Y. Polishchuk,'Active current and apparent power of three-phase power supply systems'. Technical electrodynamics, no. 6, pp. 69–72, 2018.
[35] S.Y. Polishchuk, M.Yu. Artemenko, V.M. Mykhalskyi, I.A. Shapoval, L.M. Batrak, 'Control strategy of a shunt active filter with partial attenuation of the zero-voltage component of a three-phase four-wire network', Technical electrodynamics, no. 3, pp. 12–19, 2013.
[36] M.Yu. Artemenko, L.M. Batrak, N.I. Domaskina, 'Apparent power of three-phase four-wire system in sinusoidal asymmetric mode and energy effectiveness of shunt active filters', Proceedings of IEEE XXXV International Scientific Conference "Electronics and nanotechnology", pp. 469–474, 2015.
[37] J.C. Montano, P. Salmeron, J.P. Thomas, 'Analysis of power losses for instantaneous compensation of three-phase four-wire systems', IEEE Trans. on Power Electronics, vol. 20, no. 4, pp. 901–907, July 2005.
[38] P. Salmerón, J.R. Vázquez, R.S. Herrera and S.P. Litrán, 'Apparent power and power factor in unbalanced and distorted systems. Applications in three phase load compensations', International Conference on Renewable Energies and Power Quality (ICREPQ'07). Sevilla, Spain, 28–30 Mach, 2007, vol. 1, no. 5, pp. 442–447, 2007.
[39] IEEE Standard Definitions for the Measurement of Electric Power Quantities under Sinusoidal Nonsinusoidal, Balanced or Unbalanced Conditions. IEEE Standard 1459-2010, Sept. 2010.
[40] German Standard AC Quantities: Part 2- Multi Conductor Circuits.DIN std. 40110, 2002.
[41] Artemenko M.Y. Decomposition of currents and capacities of a three-phase four-wire circuit in asymmetric non-sinusoidal mode. Proceedings of the International Scientific and Technical Conference Radio

Engineering Fields, Signals, Apparatus and Systems, March 16–22, 2015, Kyiv, pp. 22–24. 2015.

[42] Artemenko M.Y. Power supply systems and energy efficiency of power filters. Kyiv: AVERS, 2016. 216 p.

[43] Artemenko M.Y. Additional sections of power electronics. Volume 1. Matrix methods in the theory of electric circuits with valves. Kyiv: Avers, 2016. 174 p.

[44] M.Yu. Artemenko, Batrak L.M., Polishchuk S. Y. The power theory development to improve the SAF's control strategies. Lambert Academic Publishing OmniScriptum GmbH & Co. KG, Saarbrucken, Deutschland, 2017. 107 p.

[45] M.Yu. Artemenko, S.Y. Polishchuk, V.M. Mykhalskyi, I.A. Shapoval, 'Apparent Power Decompositions of Three-Phase Power Supply System to Develop Control Algorithms of Shunt Active Filter', Proceedings of the IEEE First Ukraine Conference on Electrical and Computer Engineering (UKRCON). Kyiv (Ukraine). May 29–June 2, 2017, P. 495–499. 2017.

[46] L.S. Czarnecki, 'Currents' Physical Components (CPC) concept: a fundamental of Power Theory', Przegląd Elektrotechniczny, no. 6, pp. 28–37, 2008.

[47] M.Yu. Artemenko, L.M. Batrak, 'Strategy for controlling a parallel active filter of a four-wire three-phase power supply system with multivariate proportional-vector formation of instantaneous source currents', Electronics and communication, no. 2. pp. 36–44. 2014.

[48] O.H. Hryb, Quality of electrical energy. Volume 3. Methods and means of improving the quality of electrical power. Hraf-X: Kharkiv, 2014. 292 p.

[49] IEEE Std. 1459-2010, Definitions for the measurement of electric power quantities under sinusoidal nonsinusoidal, balanced or unbalanced conditions. 2010.

[50] Yu.A. Syrotyn, O.H. Hryb, D.A. Hapon, T.S. Ierusalymova and S.V. Shvets, 'Accounting for inactive components of apparent power', Bulletin of NTU "KhPI" Hydraulic machines and hydrounits, no. 22 (1244), pp. 71–76, 2017.

[51] P. Salmerón, J.R. Vázquez, R.S. Herrera and S.P. Litrán, 'Apparent power and power factor in unbalanced and distorted systems', Applications in three phase load compensations. RE&PQJ, vol. 1, no.5, March 2007.

[52] M.Yu. Artemenko, L.M Batrak, V.M. Mykhalskyi and S.Y. Polishchuk, 'Energy performance optimization of the three phase four wire power supply system with a parallel active filter in the unbalanced sinusoidal mode', Tech. Electrodyn.. no. 2, pp. 30–37, 2015.
[53] V.M. Mykhalskyi, V.M. Sobolev, V.V. Chopyk and I. A. Shapoval, 'The minimization strategy of undesirableaneous power components with different topologies of shunt active filter', Tech. Electrodyn., no. 1, pp. 41–50, 2014.
[54] Pedro A. Blasco, Rafael Montoya-Mira, José M. Diez and Rafael Montoya, 'An Alternate Representation of the Vector of Apparent Power and Unbalanced Power in Three-Phase Electrical Systems', Appl. Sci., vol. 10, pp. 3756, 2020.
[55] A.S. Vijay, Suryanarayana Doolla, Mukul C. Chandorkar, 'Unbalance mitigation strategies in microgrids', IET Power Electron., vol. 13, iss. 9, pp. 1687-1710, 2020.
[56] L.S. Czarnecki, 'Currents' Physical Components (CPC) in Circuits with Nonsinusoidal Voltages and Currents Part 2: Three-Phase Three-Wire Linear Circuits', Electrical Power Quality and Utilization Journal, vol. XII, no. 1, pp. 3–13, 2006.
[57] Augusto Matheus dos Santos Alonso, Danilo Iglesias Brandao, Elisabetta Tedeschi and Fernando Pinhabel Marafão, 'Distributed selective harmonic mitigation and decoupled unbalance compensation by coordinated inverters in three-phase four-wire low-voltage networks', Electric Power Systems Research. vol. 186, pp. 1–14, Sept. 2020.
[58] Guodong Liu, Thomas B. Ollis, Bailu Xiao, Xiaohu Zhang and Kevin Tomsovic, 'Distributed energy management for community microgrids considering phase balancing and peak shaving', IET Gener. Transm. Distrib., vol. 13, iss. 9, pp. 1612-1620, 2019.
[59] V.S. Boiko, V.V. Boiko, Yu.F. Vydolob, Theoretical Fundamentals of Electrical Engineering: Tutorial. Steady-state modes of linear electric circles with lumped parameters. Vol. 1. Kyiv: Polytechnic Publishing House, 2004, 232 p.
[60] M.Yu. Artemenko, Yu.V. Kutafin, I.A. Shapoval, V.V. Chopyk, V.M. Mykhalskyi, S.Y. Polishchuk, 'Identification of unbalanced power in a three-phase three-wire power system under linear loading', Proceedings of the IEEE 8th International Conference on Energy Smart Systems (ESS). Kyiv (Ukraine). 2022.

5

Evolution of Microgrid, Hierarchical Architecture of Microgrid

Rajan Singaravel, Hemachander Allamsetty, and Sankar Natarajan

Department of Electrical and Electronics Engineering, National Institute of Technology Puducherry, India
E-mail: rajan.singaravel@nitpy.ac.in; hemachander.a@nitpy.ac.in; naseee2591@gmail.com

Abstract

This chapter explores the evolution and hierarchical architecture of microgrids (MGs), emphasizing their significance in the context of distributed generation and electric vehicles integration. The demand for sustainable and resilient power systems is increasing, and microgrids are emerging as essential solutions that facilitate the integration of renewable energy sources and improve grid reliability. The chapter begins with an introduction to microgrids, outlining the key drivers behind their development, such as the need for energy independence, environmental sustainability, and improved energy management. A comprehensive classification of microgrids is provided, categorized based on market segments and system topology, which highlights their diverse applications and operational frameworks. In addition to the classification, the chapter delves into the critical components of microgrids, including loads, distributed energy resources (DERs), and distributed energy storage (DES). These elements play a vital role in managing energy flow and optimizing system performance. A significant focus of the chapter is on the microgrid control system, which is organized into hierarchical levels: primary, secondary, and tertiary control. The chapter elaborates on specialized control strategies that include voltage regulation, converter control, and energy management techniques. It also discusses the implementation of these controls to ensure stable and efficient operation of microgrids. By outlining the evolution, structure, and control mechanisms of microgrids, this chapter contributes to the understanding of how distributed generation can be effectively managed to create a more resilient and sustainable energy future.

Keywords: Microgrids, distributed energy resources, distributed energy storage, hierarchical control, loads.

5.1 Introduction

The conventional power grid is an interconnected system that transports energy from generators to consumers. It also includes high tension transmission networks that transport electricity from distant sources to demand centers and distribution lines that link individual customers. The major share of the power production is delivered by centralized power-generating stations. It is typically required to be situated away from the load center because it generates the majority of the power and a large portion of the land in the load center is not readily available. The majority of them generate steam to power the turbine generator using massive fossil-fired gas or coal boilers [1]. Depending on the availability of that enormous tank and other infrastructure, large hydro may occasionally be utilized. These factories feature numerous power plant instruments and numerous plants that need a lot of water and cooling, necessitating the expensive management of substantial infrastructures. It requires huge infrastructure and is also susceptible to unreliability and instability under many foreseeable events.

The conventional electric grid is a unidirectional network, which indicates the flow of energy from generators to substations, via transmission lines, and finally to domestic or industrial customer [2]. It should also be mentioned that the majority of the infrastructure, including the lines and equipment of the traditional power system was established many years ago. They require significant investments; therefore, their provisioning usually takes years. Because of their age, these grid components require periodic upkeep and supervision in order to keep the system effective. Additionally, fossil fuels are continually being used up which makes them more and more expensive based on market prices [3, 4]. The conventional grid faces challenges such as load balancing, delivering electricity to remote areas, estimating future needs for the capacity of the delivery infrastructure (the transmission and distribution lines), increased demand for power during peak demand, and frequent failures and blackouts caused by weather and natural disasters.

In recent years, distributed generation (DG) of electric power has gained popularity as a modern solution due to its numerous benefits, including high efficiency and environmental protection, reduction of transmission and distribution losses, support for the local power grid, and improved system stability. DG units are frequently integrated with renewable energy technologies such as wind, PV, biogas, hydro, etc. DGs can cause as many problems as they can solve. A system approach that views generation and associated loads as a subsystem or "Microgrid" is a possible way to realize the emerging potential of distributed generation [5].

5.2 Microgrids

According to IEEE standards, a microgrid (MG) is a cluster of interconnected loads and distributed energy resources with clearly defined limitations which can work as a single controllable unit with respect to the utility grid and is able to operate in both grid-tied and islanded modes [6] as shown in Figure 5.1. Microgrids are defined by their function, not their size. Power generation, distribution, transmission, and control characteristics of microgrids are presumptively similar to those of a traditional grid structure. Because a microgrid is installed close to the load sites, it differs from a conventional grid in terms of the separation between cycles of power generation and consumption [7]. Further, distributed generation facilities like combined heat and power (CHP) and renewable energy sources can be integrated with microgrids. Microgrids stand out from other grid upgrading efforts due to two key characteristics:

1. A microgrid is a network of fixed-capacity generation and load centers.
2. A microgrid can function as a whole in both grid-tied and autonomous modes.

A MG operates normally in grid-connected mode when there are no power quality issues on the utility grid to which it is connected. Depending on the overall amount of electricity generated by the nearby DG, the MG in grid-tied mode may either import or export power to the utility grid or serve the full local load. The market policies may also allow the MG to maintain a predetermined load mix. When a power quality issue on the main grid arises, the MG might disconnect from it when it is in islanded mode [8]. The benefits and significance of the MG are described here. Natural disasters, blackouts, and other events do not disrupt the regular electricity supply to the MG area. If MG can meet the local demand, the cost of transmission and distribution (T&D) losses of the electricity system is reduced [9]. The installation of

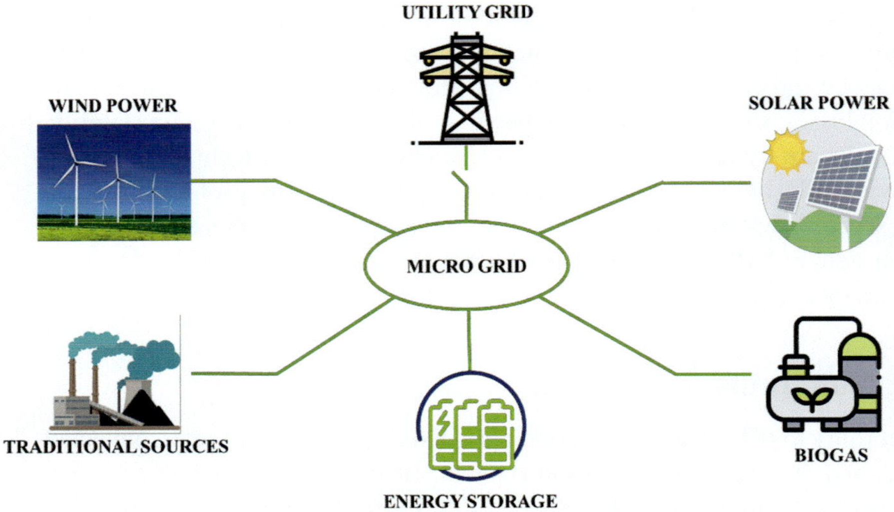

Figure 5.1 Simple diagram of a typical microgrid.

MG can delay the cost of T&D growth despite the rising demand for power. MG generates its energy mostly through environmentally friendly, renewable methods that produce nearly no emissions. The installation of an MG significantly improves the customer's ability to rely on and get high-quality power. By installing a MG with the available renewable energy sources in rural areas and by combining heat and power generating in cities, the cost per unit of energy is significantly decreased [8].

5.3 Drivers of Microgrid

There are two main groups of criteria influencing microgrid deployment in areas with an established electrical grid infrastructure. Value drivers, technology or policy drivers as follows:

5.3.1 Value drivers

I. **Resiliency/reliability:** Contingent electric service capacity is offered by microgrids in the event that the grid is down. They can also be used as an alternative to more conventional methods of ensuring power quality, including the deployment of a sizable uninterruptible power supply in a data center.

II. **Energy/capacity/grid services (energy cost savings):** Microgrids can capture economic value by combining local resources to offer the electrical grid reserve and regulation services (i.e., real and reactive power). Examples include energy efficiency, local generation sources, demand response capability (which has dramatically expanded since 2010), and load management control. These resources can be used by microgrids to take advantage of dynamic pricing to reduce consumer prices through energy arbitrage or to give the grid the energy or capacity it needs.

III. **DER integration/aggregation:** In order to provide ancillary services, for instance, microgrids can group and manage a number of installed DERs and run them as a single entity from the utility's or grid operator's point of view. This aggregation function can help with grid planning and operation.

IV. **Emissions reduction:** When compared to conventional generation assets, a microgrid's renewable (or higher efficiency) DERs can reduce carbon emissions. This is especially true in several scenarios.

V. **Energy:** Microgrids provide distant communities or entities of great national interest with the ability to remain off the grid for extended periods of time and a reliable power supply. Greater cybersecurity requirements for the former can be met by the enhanced reliability. Renewable energy sources generally reduce the danger posed by erratic or interrupted petroleum supply.

VI. **Investment alternative/deferral:** Utilities can postpone distribution asset upgrades by utilizing local resources inside a microgrid to lower grid energy use during peak hours. These local resources can also offer an alternative investment plan to support resilience and dependability advancements instead of traditional tactics (e.g., supplanting the need for a redundant transmission line).

5.3.2 Technology/policy drivers

I. **Technology and cost:** The cost of renewable energy sources, such as solar PV and wind has drastically fallen while performance has increased during the last 10 years. Due to technological advancements and economies of scale, the cost of conventional DG (such as internal combustion engines and compact combustion turbines) and energy storage has also decreased. Greater potential for waste heat capture in CHP systems are also being created by energy conversion to electricity close to end users. All of these values are capturable by microgrids.

II. **Advanced controls and power electronics:** Improved inverters result from new power electronics technology including power conditioning and control (needed for direct current sources, such as solar PV and battery energy storage that are becoming prevalent in microgrid installations). These technologies are enabling better DER integration and control by offering grid support functionalities.

III. **Public policy:** DG that offers more efficiency, reduced emissions, and higher power system security is favored by current public policies and incentives, which include renewable energy tax credits, renewable portfolio standards, grid modernization funds, emissions limitations, and net metering.

IV. **Energy user awareness:** The readiness to take into account onsite generation possibilities is growing as energy users become more aware of alternate power sources and the economics that go along with them.

5.4 Types of Microgrids

The microgrids can be classified based on system architecture and market segments [10]. Subsets of both classifications are discussed below in detail.

5.4.1 Classification of microgrids based on system topology

I. **DC microgrid:** The DC microgrid has become a modern prodigy in the power system as a result of technological developments and PECs. The DC microgrids are gaining significance because of their flexibility and ability to naturally interface with batteries and RES like PV, as well as their efficiency due to little power conversion as shown in Figure 5.2. Additionally, more extensive research in areas like RES and DC microgrid has demonstrated the benefits of the latter, advancing the technology toward real-time use.

II. **AC microgrid:** This traditional type of microgrid is most frequently used. A variety of renewables, including solar PV, wind, and fuel cells are connected to the existing utility grids or big power networks as shown in Figure 5.3. Due to the established network, AC microgrid requires little change and alters the topology just slightly. Due to their capacity to expand the distribution network with minimum losses, this system is frequently used with medium/low voltage levels. Although connecting a microgrid to a conventional architecture is advantageous, there are several drawbacks as well, including system stability, power quality, a lack of reactive power, and DER synchronization.

III. **Hybrid AC–DC microgrid:** An AC microgrid and a DC microgrid are connected together to form a hybrid microgrid. Although the architecture of hybrid microgrid is complex, it has advantages over either microgrid, including flexibility, improved efficiency, dependability, and cost-effective operation. Due to the AC microgrid architecture, the hybrid topology makes it easy to connect to the traditional grid, whereas

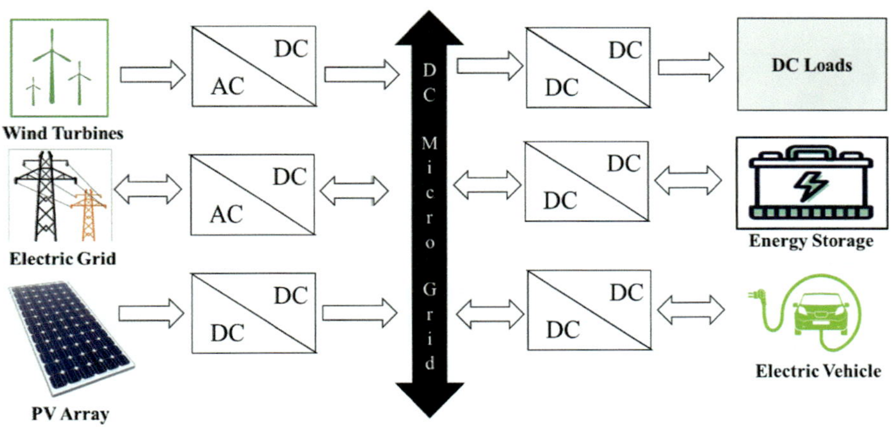

Figure 5.2 DC microgrid.

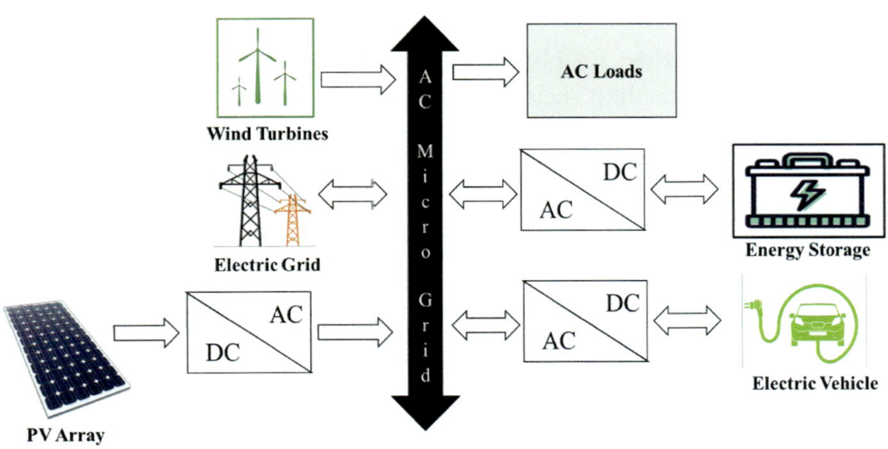

Figure 5.3 AC microgrid.

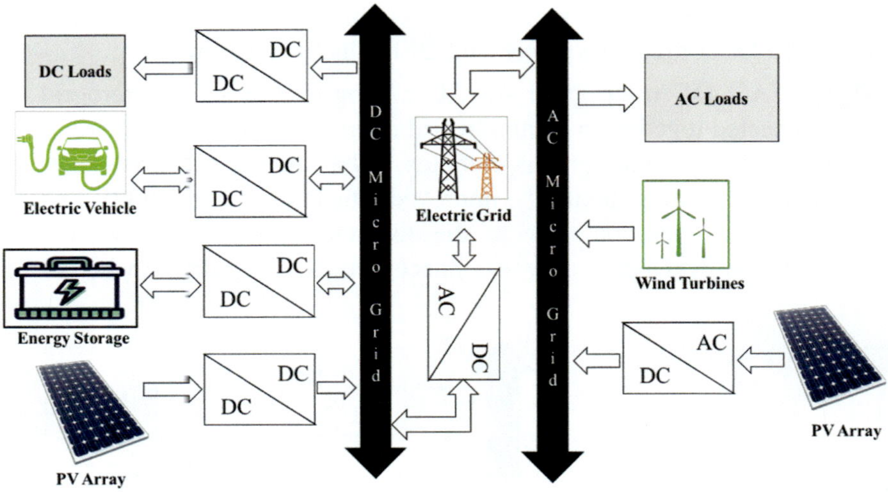

Figure 5.4 Hybrid AC–DC microgrid.

the DC microgrid architecture makes it possible to combine DC-based RES, DERs, ESSs, and loads with little modification to the current system as shown in Figure 5.4. The hybrid microgrid topology significantly decreases the number of power electronic converters required, which is followed by a decrease in unnecessary losses caused by power conversion.

5.4.2 Classification of microgrids based on market segments

I. **Campus microgrids:** Enabling onsite generation, in particular with the combination of cooling, heat, and power application with variety of loads on a premises maintained as a single entity with capacity ranging from up to 40 MW. A campus microgrid can offer the benefit of peak load reduction.

II. **Military microgrids:** It has the means to utilize dependable fuel and electricity supplies to secure and deliver enough energy to support essential operations even during a protracted failure of the nearby electrical grid [11].

III. **Residential microgrids:** A residential microgrid is made up of a sophisticated control system that integrates the customer's electrical needs, controls distributed resources like solar PV and energy storage, and works with the transmission and distribution networks for electricity.

A residential microgrid lessens a customer's reliance on centralized electricity supply by providing emergency power to crucial circuits during power outages.

IV. **Remote and rural microgrids:** Off-grid microgrids are physically isolated from the utility grid and always operate in islanded mode due to a lack of available and inexpensive transmission and distribution (T&D) equipment nearby. Renewable energy sources, including wind and solar, often offer the microgrid operator a more practical DER option that is also more cost-effective and ecologically friendly for these remote circumstances. In addition, a lot of remote microgrids are thinking about switching from traditional generators to battery energy storage systems for backup power [12].

5.5 Components of Microgrid

The components of microgrid system are loads, distributed energy resources, distributed energy storage, and microgrid control system. This section describes the different components of microgrid.

5.5.1 Loads

Nature of the loads will have a significant impact on the configurations of microgrid. Based on the nature and priority, microgrid loads are classified as follows [13].

I. **Critical loads:** These are the loads that the microgrid must be able to handle in all of its typical operational modes. Usually, they cannot be interrupted. IT and computer power, building controls, security systems, and life-safety systems are some examples (e.g., egress lighting).

II. **Priority loads:** These are loads that may be curtailed, if necessary, but should be satisfied if possible.

III. **Controllable loads:** These are loads that can be interrupted on a regular basis or set to different load levels. Nonessential industrial process and heating, ventilation, and air conditioning (HVAC) system loads (chillers, air handlers) are examples. Building HVAC systems' operating points can be adjusted to provide controllable loading.

IV. **Interruptible loads:** These are loads that can be shed or interrupted at any time in order for generation to meet peak demand. Thermal energy storage, water pumping where sufficient tank storage is available, and battery charging in distributed energy storage (DES) devices are some examples.

V. **Diversion and dump loads:** These loads are used when there is excess generating capacity available and generation curtailment is either not possible or there is an economic benefit to using the diverted energy. A diversion load for a wind turbine that absorbs gust energy is one example.

VI. **Undefined loads:** Other loads not defined above, including loads with undefined characteristics.

5.5.2 Distributed energy resources (DERs)

Distributed energy resources can be installed at the premises of end consumers and/or electric utility facilities. DERs can reduce distance between generating station and load, losses in power transmission and carbon emissions. In addition to this, DERs can reduce transmission congestion, improve grid voltage profile and power quality, and make better use of waste heat. Further, when compared to large power generation plants and T&D facilities, DERs require less time for construction and deployment. And DERs can provide power to remote locations where electrical network facilities are either too costly to build or not available [14]. DERs can be classified into two types:

i) dispatchable units, which can be controlled by the microgrid control system;
ii) nondispatchable units, which cannot be controlled by the microgrid control system since the sources are intermittent in nature (renewables).

DERs available to develop microgrid:

- Solar photovoltaic (PV)
- Wind energy
- Biomass
- Fuel cells
- Small hydro
- Gas or diesel cogeneration/CHP

5.5.3 Distributed energy storage (DES)

The incorporation of intermittent RES within a microgrid has transformed traditional power generation resources from controllable and dispatchable to uncontrollable and nondispatchable [15]. Furthermore, stochastic EV loads make power demand unpredictable. As a result, microgrids face a significant challenge in balancing power demand and supply. Energy storage, on the other hand, becomes a viable solution for microgrids to address the unpredictability of power demand and generation. Energy storage systems are also essential in microgrids for ensuring system stability, reliability, and power quality while lowering operating costs and increasing power efficiency. The DES's primary application is to collaborate with DGs to ensure microgrid generation sufficiency.

5.5.4 Microgrid control system

Control system of the microgrid should have both real-time control and energy management functions that operate in the following situations:

a) operation in grid-tied and autonomous modes;
b) automatic transition from grid-tied and autonomous mode by providing a smooth transition to autonomous mode for MG loads during abnormal conditions and planned interruptions of the system;
c) synchronization with the main grid during the transition from autonomous mode to grid-tied mode;
d) optimizing both real and reactive power generation and consumption through proper energy management scheme.

5.6 Hierarchical Control of MG

The extensive research on microgrids (MGs) led to the evolution of various MG structures. The principal objective of the MG, as discussed is to enable intelligent power distribution from source to load through an effective communication network [16–18]. An effective control technique is essential for the efficient and stable operation of MG. This section aims to report the significant control hierarchies reported and their control strategy.

5.6.1 Conventional control hierarchy

The conventional control hierarchy includes three steps control namely, primary control, secondary control, and tertiary control [16], [19], [20]. The conventional control architecture of a microgrid is illustrated in Figure 5.5.

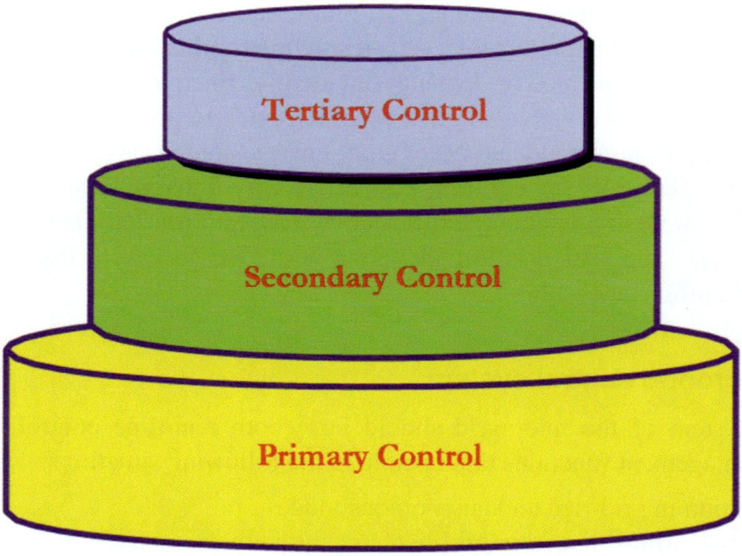

Figure 5.5 Conventional control architecture of microgrid.

5.6.1.1 Primary control

The primary control is an independent and limited control and its major objectives are to ensure [21]:

- local voltage control;
- autonomous operation of each generating unit;
- stable MG operation;
- reliability of the MG;
- effective power-sharing among the generating units.

It is noteworthy that MG has fast-changing dynamics and hence the primary control should be as fast as possible. This control depends on the local measurements and any communication to it is avoided due to reliability issues and hence is known as a local control strategy. In the grid-tied mode, generating units deliver power independent of load variations and on the other hand, in islanded mode, power is delivered based on the power-sharing and voltage control to ensure stable operation of the microgrid. This control is the primary level of the hierarchical control mainly confined to the management of voltage and stable operation. As only local control is provided, supervisory action for this control is ensured through the higher-level controllers to enhance the overall performance of the system.

5.6.1.2 Secondary control

The secondary control is a centralized control and its major objectives include [21, 22]:

- set voltage for local controller;
- restoration of voltage and frequency;
- proportional load sharing;
- compensate the deviations due to primary control;
- voltage control;
- harmonic compensation.

Unlike primary control, secondary control consists of slow control loops to facilitate individual design to ensure decoupled dynamics of the two controllers. Further, the secondary control also employs a low bandwidth communication system which enables to sense the parameters of the microgrid and direct the control output information to individual generating units. This type of control generally referred to as microgrid central controller is well established in large-scale power systems for frequency control in large-area electrical networks. Power flow management within the microgrid is completely taken care by this secondary control.

5.6.1.3 Tertiary control

The controllers discussed above control the voltage and current, however, these are not sufficient for the optimal and economical operation of the microgrid. In this regard, tertiary control is employed in either centralized or distributed means. The major objectives of tertiary control include [19, 21]:

- power and energy management;
- optimal energy storage;
- minimal operational cost.

The tertiary control being a supervisory control extends its operation beyond the microgrid to ensure optimal power flow between the microgrid and conventional grid or between the clusters of the microgrid.

To summarize, the conventional control hierarchy with their major functions are illustrated in Figure 5.6.

5.6.2 Specialized hierarchical control architecture

A specialized hierarchical architecture for DC microgrid has been detailed in [23, 24]. Similar to above, three stages of control have been defined viz., converter control level, coordination level, and energy management level. The converter control level is limited to handling the various converter

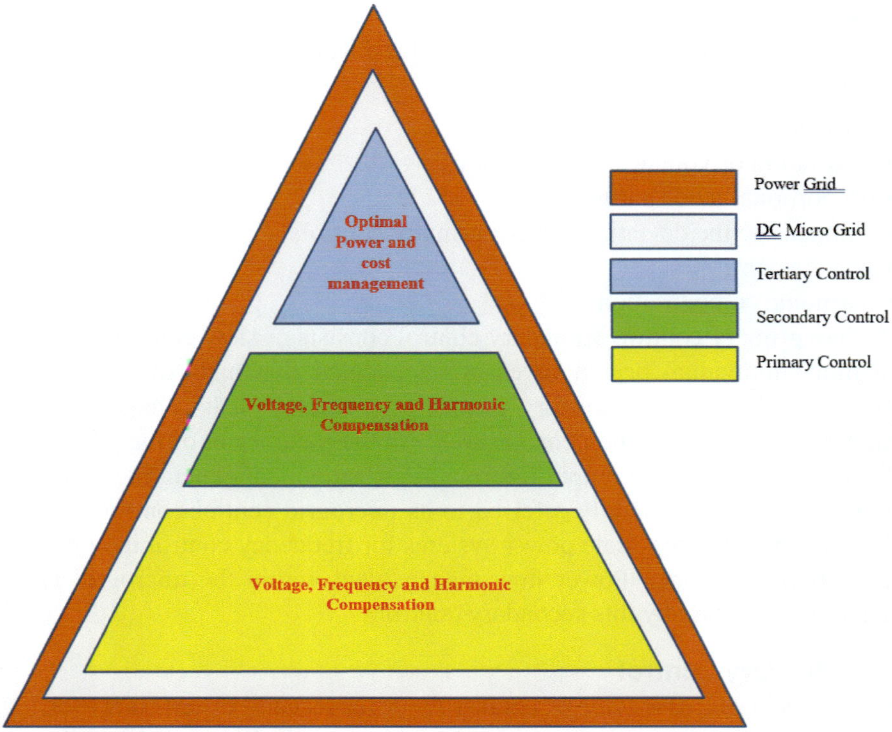

Figure 5.6 Conventional control hierarchy with their major functions.

architectures employed in the microgrid. The coordination level handles the power flow in the network using the DC voltage. The third level deals with optimizing the performance through the information network.

5.6.2.1 Converter control level

The state of the art reported numerous converter architectures for connecting the sources to the microgrid [24]. The primary task of the converter control is to analyze and control the parameters of the converters based on the command received from the higher control levels. The various converter topologies like boost, resonant, DAB have been employed for microgrid applications.

5.6.2.2 Voltage coordination level

The coordination of the converters in the microgrid through DC voltage is the major objective of the coordination control [25]. Any converter unit in a DC microgrid has power and voltage terminals. The power terminal aids in regulating the local power irrespective of the system requirements. On the other hand, the voltage terminal aids in regulating the power balance and voltage stability of the overall system. The coordinative control strategy is the key to the effective collaboration of multiple voltage terminals in the microgrid.

5.6.2.3 Energy management level

Effective and efficient control through the information network is achieved through the energy management level. When microgrid is decoupled from the grid, the tasks of energy management are also decoupled. The microgrid energy management level takes care of the microgrid energy sharing alone. The other key objectives of the energy management level include [26]:

- ✓ regulation of power quality to improve the voltage and power profiles;
- ✓ power flow optimization to ensure proper power dispatch in the grid;
- ✓ power scheduling based on meteorological conditions and load forecasts.

The stability of the system is taken care of by the energy management level with proper adjustments of the parameters in voltage coordination level.

5.7 Methods of Control

The control methods are important for proper control of power flow. There are several control methodologies reported in the state of the art [19], [27]. This section deals with the most commonly used control methods in brief.

5.7.1 Control strategies for primary control

5.7.1.1 Fuzzy logic control

It is a known fact that the fuzzy logic controller uses a set of control rules to elucidate the issues of a nonlinear system. Due to its characteristics like precision and fast, it is used as a controller for many nonlinear systems. Due to its widespread accuracy, FLC is used in the primary control for optimal voltage and current sharing among the generating units. Further, the recent studies also reported that fuzzy logic can be employed for a different range of control frameworks without prior knowledge of the historic data which made FLC to be employed in the primary control. Few works also reported the usage of FLC for energy sharing between the photovoltaic and battery, battery and ultracapacitor and many more for optimal power flow. Moreover, the smart method of FLC has made it penetrate into the applications like state of charge management to avoid overcharge and underdischarge of battery which results in improved battery performance and life.

5.7.1.2 DC bus signaling

This is an economical distributed control strategy that regulates the voltage of the DC bus and also performs the source scheduling in the system without the means of the communication network. The source scheduling is performed with the consideration of the source and the limits on charging and discharging of storage. In this strategy, bus voltage and DC bus behave as an information carrier and communication link respectively. The DC bus signaling is primarily employed to maintain the stability of bus voltage and effective power management among the converters under all generating and loading scenarios [28]. In this method, an interface is established among the converters, local buses, energy storage elements and utility grid if required. This method is the most extensively implemented distributed control approach due to the advantages like improved steady-state and dynamic performance, enhanced reliability; however, deviation in the voltage is high which is the major limitation.

5.7.1.3 Droop control

The droop controller is a local controller which offers a technique for non-feedback paralleling schemes without any secondary circuit for sharing the information [29]. Further, this also offers significantly reasonable voltage and current regulations. The basic droop control is used in linear approach only; however, the recent literature reported that it can be used in a nonlinear approach too. The performance of the linear droop control is adversely affected by the line impedance. To overcome this effect of line impedance, a nonlinear approach has been adopted. The nonlinear characteristics hinder the trade-off between voltage regulation and current sharing [30]. Further, for effective current sharing and voltage regulation, a droop coefficient has been introduced to regulate the behavior of droop control.

5.7.2 Secondary control scheme

The secondary control level plays an important role in the DC microgrid for bridging the primary and tertiary control. Similar to the primary control, its operation is limited to the microgrid only. The major role of the secondary control lies in improving the voltage regulation by providing a set point to the local controller and initiating proper current sharing among the converters, thereby improving the power-sharing capacity and power quality of the microgrid [31]. Further, this also is responsible for optimal power flow within the microgrid and also in the cluster of the microgrid. Further, recent reports validate the effectiveness of the secondary control for control of DC bus voltage and state of charge of the battery in a DC microgrid cluster.

The secondary controller can be adopted in a centralized, decentralized, and distributed approach [32]. In a centralized approach, a central controller is linked with sources and loads of DC microgrid through communication links. It measures the common bus voltage at the load point where all loads are connected and then directs the set point to the voltage controller. The centralized approach of the secondary control offers highly accurate, flexible and suitable control action. However, the major limitation lies in the large communication network, resulting in decreased reliability of the system. On the other hand, due to very less communication requirements, decentralized control is comparatively less complex. The decentralized control performs local regulation based on the available local variables of separate units by means of different strategies like the droop-based feedforward control for terminal voltage reference, proportional controller for cable impedance balancing, etc. Simplified implementation of the decentralized control improves the modularity and reliability of the system. Most of the various decentralized control strategies reported are not accurate in rapidly changing load scenarios. In this regard, the reference output voltage is generated by means of a distributed consensus algorithm using the average values of current and voltage in distributed control. Further, using separate average voltage and current controllers enhances the converter current sharing in distributed control [33]. The secondary control is responsible to enhance the power-sharing capability of the clustered microgrid and hence it is significant to devise a proper secondary control scheme for microgrid.

5.7.3 Tertiary control

The tertiary control is an additional level that aids in achieving the economic and optimal operation of the microgrid and can be implemented in either a centralized or a distributed way. The three essential requirements of the control scheme being adopted in the tertiary control are reliability, economic and optimal state of the system [34]. The economic and optimal state is highly taken care of by the tertiary control. It is noteworthy that any system desires to operate at or near to the optimal state and the tertiary controller plays an important role in this aspect. Most often, for the selection of the

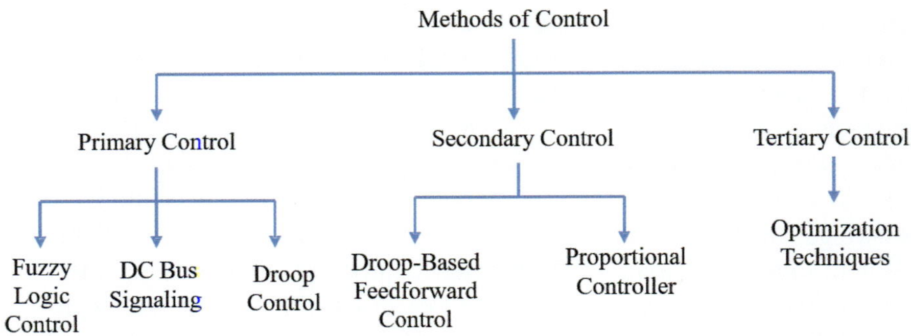

Figure 5.7 Classification of control algorithm.

desired optimal state, optimization techniques like genetic algorithm, particle swarm optimization, etc. are used [27]. These optimization techniques use the modified input and reduce the burden of complex computational efforts for determining the optimal state of operation.

The various control algorithms used in the conventional hierarchical control is summarized in Figure 5.7.

5.8 Conclusion

This chapter discusses superlatively about the classification of microgrids and their hierarchal architecture. Various components of microgrids like distributed energy resources, energy storage and loads are discussed in detail. Different hierarchy of microgrid control system is explained for both conventional and specialized control architectures. Further, at the end of the chapter, various control methods required for microgrid are also discussed. Hence, this chapter consolidates the key information about the microgrid which will help to understand the inevitability of microgrid in the near future.

References

[1] Nihal Kularatna, Kosala Gunawardane, “1 - Modern electrical power system and the role of distributed generation”, Editor(s): Nihal Kularatna, Kosala Gunawardane, Energy Storage Devices for Renewable Energy-Based Systems (Second Edition), Academic Press, 2021, Pages 1-35, ISBN 9780128207789.

[2] S. Khoussi, A. Mattas, “11 - A Brief Introduction to Smart Grid Safety and Security”, Editor(s): Edward Griffor, Handbook of System Safety and Security, Syngress, 2017, Pages 225-252, ISBN 9780128037737.

[3] H. Farhangi, "The path of the smart grid", IEEE Power Energy Mag. 8 (1) (2010) 18-28.

[4] S.M. Kaplan, "Electric Power Transmission: Background and Policy Issues", Congressional Research Service, Washington, DC, 2009.

[5] R. H. Lasseter, "MicroGrids," in 2002 IEEE Power Engineering Society Winter Meeting, New York, NY, 2002, pp. 305-308.

[6] Ton DT, Smith MA. The US department of energy's microgrid initiative. Electr J 2012;25: 84–94.

[7] Hossain, Eklas & Kabalcı, Ersan & Bayindir, R. & Perez, R. (2014). "A comprehensive study on microgrid technology", International Journal of Renewable Energy Research. 4. 1094-1104.

[8] Banerji et al., "Microgrid: A review," 2013 IEEE Global Humanitarian Technology Conference: South Asia Satellite (GHTC-SAS), 2013, pp. 27-35, doi:10.1109/GHTC-SAS.2013.6629883.

[9] P. Chiredeja and R. Ramakumar, "An approach to quantify the technical benefits of distributed generation," IEEE Trans. Energy Conversation, vol.19, no. 4, pp. 746-773, Dec. 2004

[10] Ayush Mittal, Aryan Rajput, Kamya Johar, Ritu Kandari, "Chapter 1 - Microgrids, their types, and applications", Editor(s): Josep M. Guerrero, Ritu Kandari, Microgrids. Academic Press, 2022, Pages 3-40, ISBN 9780323854634.

[11] Adam Hirsch, Yael Parag, Josep Guerrero, "Microgrids: A review of technologies, key drivers, and outstanding issues", Renewable and Sustainable Energy Reviews, Volume 90, 2018, Pages 402-411, ISSN 1364-0321.

[12] Rahmat Khezri, Amin Mahmoudi, Mohammad Hassan Khooban,"1 - Microgrids planning for residential electrification in rural areas",Editor(s): Sanjeevikumar Padmanaban, C. Sharmeela, P. Sivaraman, Jens Bo Holm-Nielsen, Residential Microgrids and Rural Electrifications, Academic Press, 2022, Pages 1-25, ISBN 9780323901772.

[13] Zhang L. Sun M. "Research on the type of load of accessing to micro-grid based on reliability", In Proceedings2nd International conference on electrical, computer engineering and electronics, Jinan, China. pp. 1665–70.

[14] Vijay K. Sood, Haytham Abdelgawad, "Chapter 1 - Microgrids architectures", Editor(s): Rajeev Kumar Chauhan, Kalpana Chauhan, Distributed Energy Resources in Microgrids, Academic Press, 2019, Pages 1-31, ISBN 9780128177747.

[15] Nazaripouya, Hamidreza & Chung, Yu-Wei & Akhil, Abbas. (2019). "Energy Storage in Microgrids: Challenges, Applications and Research Need", International Journal of Energy and Smart Grid. 3. 60-70. 10.23884/IJESG.2018.3.2.02.

[16] T. L. Vandoorn, J. C. Vasquez, J. De Kooning, J. M. Guerrero, and L. Vandevelde, "Microgrids: Hierarchical control and an overview of the control and reserve management strategies," *IEEE Ind. Electron. Mag.*, vol. 7, no. 4, pp. 42–55, 2013, doi:10.1109/MIE.2013.2279306.

[17] J. Kumar, A. Agarwal, and V. Agarwal, "A review on overall control of DC microgrids," *J. Energy Storage*, vol. 21, no. April 2018, pp. 113–138, 2019, doi:10.1016/j.est.2018.11.013.

[18] M. Llic, "Micro-grid operation and control Challenges and expected functionalities," *IEEE Electrif. Mag.*, pp. 65–74, 2021, [Online].

[19] A. Bidram and A. Davoudi, "Hierarchical structure of microgrids control system," *IEEE Trans. Smart Grid*, vol. 3, no. 4, pp. 1963–1976, 2012, doi:10.1109/TSG.2012.2197425.

[20] C. N. Papadimitriou, E. I. Zountouridou, and N. D. Hatziargyriou, "Review of hierarchical control in DC microgrids," *Electr. Power Syst. Res.*, vol. 122, pp. 159–167, 2015, doi:10.1016/j.epsr.2015.01.006.

[21] J. M. Guerrero, J. C. Vasquez, J. Matas, L. G. De Vicuña, and M. Castilla, "Hierarchical control of droop-controlled AC and DC microgrids - A general approach toward standardization," *IEEE Trans. Ind. Electron.*, vol. 58, no. 1, pp. 158–172, 2011, doi:10.1109/TIE.2010.206 6534.

[22] P. Nahata, A. La Bella, R. Scattolini, and G. Ferrari-Trecate, "Hierarchical Control in Islanded DC Microgrids with Flexible Structures," *IEEE Trans. Control Syst. Technol.*, vol. 29, no. 6, pp. 2379–2392, 2021, doi:10.1109/TCST.2020.3038495.

[23] Y. Gu, H. Yang, W. Sun, Y. Chi, W. Li, and X. He, "Hierarchical control of DC microgrids robustness and smartness," *CSEE J. Power Energy Syst.*, vol. 6, no. 2, pp. 384–393, 2020, doi:10.17775/CSEEJPES.2017 .00920.

[24] C. M. Lai, C. T. Pan, and M. C. Cheng, "High-efficiency modular high step-up interleaved boost converter for dc-microgrid applications," *IEEE Trans. Ind. Appl.*, vol. 48, no. 1, pp. 161–171, 2012, doi:10.1109/TIA. 2011.2175473.

[25] D. Chen, L. Xu, and L. Yao, "DC voltage variation based autonomous control of DC microgrids," *IEEE Trans. Power Deliv.*, vol. 28, no. 2, pp. 637–648, 2013, doi:10.1109/TPWRD.2013.2241083.

[26] D. Zhang, S. Li, P. Zeng, and C. Zang, "Optimal microgrid control and power-flow study with different bidding policies by using powerworld simulator," *IEEE Trans. Sustain. Energy*, vol. 5, no. 1, pp. 282–292, 2014, doi:10.1109/TSTE.2013.2281811.

[27] A. Abhishek, A. Ranjan, S. Devassy, B. Kumar Verma, S. K. Ram, and A. K. Dhakar, "Review of hierarchical control strategies for DC microgrid," *IET Renew. Power Gener.*, vol. 14, no. 10, pp. 1631–1640, 2020, doi:10.1049/iet-rpg.2019.1136.

[28] D. Wu, F. Tang, T. Dragicevic, J. M. Guerrero, and J. C. Vasquez, "Coordinated control based on bus-signaling and virtual inertia for Islanded DC Microgrids," *IEEE Trans. Smart Grid*, vol. 6, no. 6, pp. 2627–2638, 2015, doi:10.1109/TSG.2014.2387357.

[29] E. Planas, A. Gil-De-Muro, J. Andreu, I. Kortabarria, and I. M. De Alegría, "Design and implementation of a droop control in d-q frame for islanded microgrids," *IET Renew. Power Gener.*, vol. 7, no. 5, pp. 458–474, 2013, doi:10.1049/iet-rpg.2012.0319.

[30] A. C. Z. De Souza, M. Santos, M. Castilla, J. Miret, L. G. De Vicuña, and D. Marujo, "Voltage security in AC microgrids: A power flow-based approach considering droopcontrolled inverters," *IET Renew. Power Gener.*, vol. 9, no. 8, pp. 954–960, 2015, doi:10.1049/iet-rpg.2014.0406.

[31] M. K. Senapati, C. Pradhan, S. R. Samantaray, and P. K. Nayak, "Improved power management control strategy for renewable energy-based DC micro-grid with energy storage integration," *IET Gener. Transm. Distrib.*, vol. 13, no. 6, pp. 838–849, 2019, doi:10.1049/iet-gtd.2018.5019.

[32] L. C. Link, S. Peyghami-akhuleh, and H. Mokhtari, "Distributed Secondary Control in DC Microgrids with Low-Bandwidth Communication Link," *2016 7th Power ...*, no. Pedstc, pp. 16–18, 2016, [Online].

[33] S. Sahoo and S. Mishra, "A Distributed Finite-Time Secondary Average Voltage Regulation and Current Sharing Controller for DC Microgrids," *IEEE Trans. Smart Grid*, vol. 10, no. 1, pp. 282–292, 2019, doi:10.1109/TSG.2017.2737938.

[34] L. Meng *et al.*, "Review on Control of DC Microgrids and Multiple Microgrid Clusters," *IEEE J. Emerg. Sel. Top. Power Electron.*, vol. 5, no. 3, pp. 928–948, 2017, doi:10.1109/JESTPE.2017.2690219.

6

Multiport DC–DC Converter Fed Multilevel Inverter for Standalone Photovoltaic Systems

Hemachander Allamsetty[1], Lalit Kumar Sahu[2], and Praneeth Kumar Pedapati[1]

[1]Department of Electrical & Electronics Engineering, NIT Puducherry, India
[2]Department of Electrical Engineering, NIT Raipur, India
E-mail: hemachander.a@nitpy.ac.in; lkumar.ele@nitrr.ac.in; ee21d1008@nitpy.ac.in

Abstract

Increasing demand for inverters in photovoltaic standalone applications for residential purposes has been a greater concern for many years. The major concern with the existing systems is their inability to extract maximum power under all conditions, especially, partial shading. The performance of the module is degraded by partial shading. In addition, it also affects the efficiency of the complete system. The literature has reported many solutions for mitigating the effect of partial shading. However, the solutions reported suffering either from complex computational efforts or a high component count. In this regard, a multiport converter structure is discussed in this chapter to extract maximum power from individual modules even during partial shading conditions. The multiport converter described not only handles partial shading conditions but also effectively handles the module mismatch conditions. Further, the multiport converter has an inherent fault-tolerant capability which results in increased reliability of the overall system. The other key features include modular structure, simple design, less component count, and a single inductor which makes the system more economic and decreases the volume of the overall system. To make the output of the multiport converter feed the residential loads, it is connected to a multilevel inverter structure to provide a superior voltage response and reduce the filter requirements of the system. Further, investigations have been carried out to mark the system as a transformerless inverter system by gaging the leakage current and common-mode voltage of the system. The proposed system has been validated for different operating scenarios like partial shading and dynamic loading conditions in MATLAB/Simulink environment. The approach for designing the controller has been described in detail. The system can operate with the respective MPP and concurrently meet the load mandate while regulating the required voltage of grid owing to the converter and controller architecture presented. The routine of the controller under various operating scenarios have also been assessed in MATLAB/Simulink environment.

Keywords: Multilevel Inverter, Maximum Power Point Tracking, DC-DC Converters, PV Panels, Efficiency Analysis.

6.1 Introduction

In recent years, transformerless PV inverters for residential applications have drawn the attention of industry and academia. Several inverter configurations have been presented in the literature that are transformerless. The transformerless inverter depicted in Figure 6.1 (a) is a centralized single-stage which has gained popularity among the available structures. These structures are formed by a single or multiple PV strings connected to the grid or load [1–2]. Replacement of single-stage conversion with two-stage conversion is another type of transformerless inverter layout that has gained traction as shown in Figure 6.1 (b) [3–4]. Both of these structures are highly efficient and cost-effective. However, they lack the ability to track individual panels' maximum power point (MPPT), resulting in a significant loss in power extraction, particularly during partial shading. A series connection of several panels also results in a high DC link voltage, leading to a dangerous arc fault. Single panel-based inverters have been presented to overcome these difficulties and have gained favor due to their superior MPPT capabilities. These series-connected structures provide the option of increasing the voltage to meet the requirements. In these structures, each panel is coupled to a converter that is nonisolated, which is then connected in series. Finally, as depicted in Figure 6.1 (c), the DC–DC converters are placed in series prior to centralized inverter [5–6]. This system is quite effective in tracking the MPPT of individual panels; however, as the number of DC–DC converters increases,

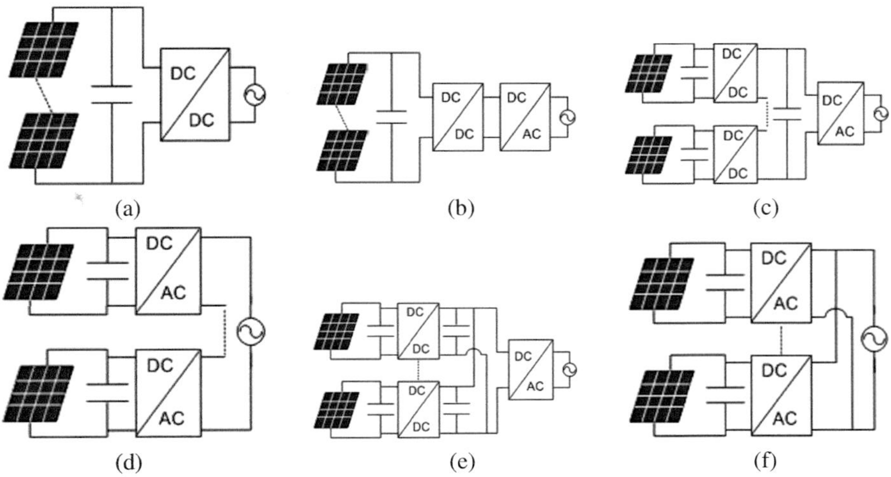

Figure 6.1 Standard 1-phase transformerless converters: (a) one stage, (b) dual stage, (c) dual stage series configuration, (d) one stage series configuration, (e) dual stage parallel configuration, and (f) one stage parallel configuration.

so does the component count, resulting in higher costs and lower dependability. As indicated in Figure 6.1 (c), another series-connected structure in each panel is linked to an inverter, which is subsequently coupled in series to the load or grid [7–8]. In comparison to the previous version of the series-based structure, this uses an H-bridge inverter and has various advantages such as a simple configuration and a lower component count. Still, because of buck topology, the component count and the quantity of panels connected in series is limited, making it difficult for standalone applications.

The parallel type structure is another solution reported in the state of the art. Each panel is connected to an individual MPPT converter and then connected in parallel, similar to a series panel-based approach. Finally, as illustrated in Figure 6.1 (e) these parallel connected converters are coupled to a centralized inverter [9–10]. These efficiently extract the maximum power from each panel, however, like series-connected panels, they have a high component count. Parallel isolated DC–AC inverters, also known as microinverters, are another parallel-type solution that has proven successful in the market. Figure 6.1 (f) depicts this structure [11–12]. These have been proven to extract maximum power in a simple and adaptable manner under all conditions. However, compared to transformerless string inverters, this necessitates a high step-up isolated converter, which increases the magnetic components, resulting in more cost and lower efficiency.

The literature reported either fail to extract MPPT under all conditions, or result in increased component count, and increased magnetic components. In this connection, this chapter proposes a multiple input structure for extracting the maximum power from each panel, which is connected to a conventional boost converter for increasing the voltage level, and then fed to a T-type multilevel inverter. This configuration effectively draws maximum power from each panel, surpassing all of the limitations mentioned.

6.2 Operation of the Proposed Configuration

The proposed PV inverter configuration depicted in Figure 6.2 employs a multiple input structure cascaded to a boost converter connected to a modified T-type multilevel inverter. Individual PV panels are connected to a switch whose duty is controlled by the respective MPPT algorithm, ensuring maximum power extraction from each panel under all atmospheric conditions and reducing power loss under partial shading. This voltage is then fed through a boost converter, which boosts the voltage level even further to satisfy the load requirement, before being delivered to a DC–AC inverter. Figure 6.3 depicts the PV panels' switching strategy when coupled to the dual input configuration.

The PV panels are independent in their operation, as illustrated in Figure 6.3, and the on-time of the switches S_{pv1} and S_{pv2} is governed by their individual MPPT algorithms. This assures that the PV panels harvest maximum power for all conditions, including partial shading on one panel having no effect on the operational point of the other. This is accomplished by a switching strategy based on time multiplexing.

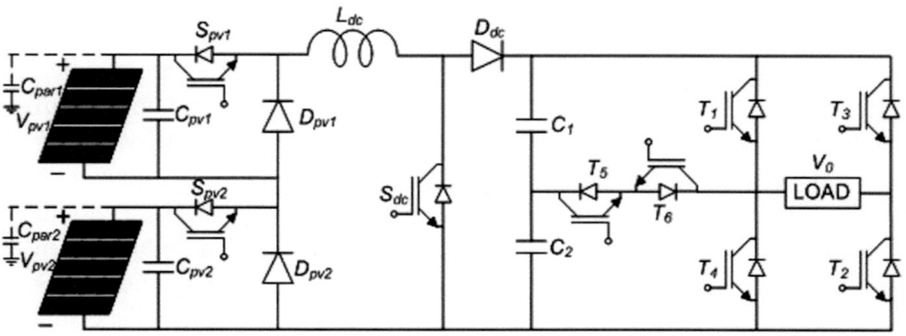

Figure 6.2 Structure of the proposed dual input dual stage transformerless multilevel PV inverter.

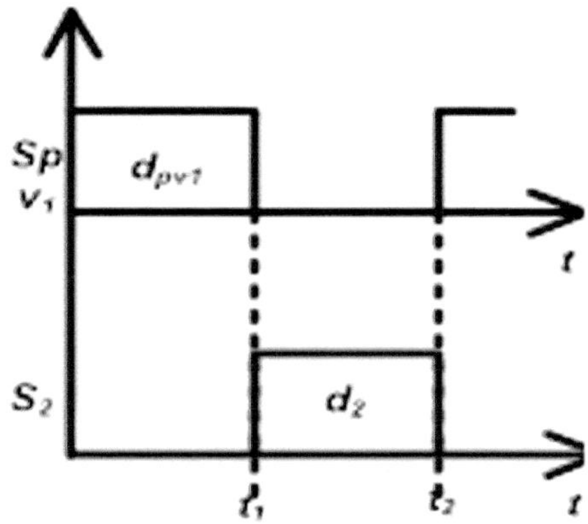

Figure 6.3 Switching scheme for dual input structure.

As previously indicated, the switches S_{pv1} and S_{pv2} assist in operating the panels at their respective MPPTs, while switch S_{dc} boosts the voltage level to the needed level. The boost converter's output voltage is given below in eqn (6.1) [13]

$$V_{\rm dc} = \frac{V_{pv1}d_1 + V_{pv2}d_2}{1 - d_{\rm dc}}, \tag{6.1}$$

where V_{dc} is the capacitor's voltage C_1 and C_2 (i.e., after boosting the generated voltage), d_1 and d_2 are the respective duty ratios of switches S_{pv1} and S_{pv2} obtained by their respective MPPT algorithms (P&O method), and d_{dc} is the duty ratio of switch S_{dc}.

The voltage V_{dc} is the input to the T-type multilevel inverter (MLI). In a traditional T-type MLI, the capacitors are required to divide the input voltage, however, the proposed topology uses the capacitors for both DC link and voltage division. This reduces the requirement of a DC link capacitor, resulting in improved reliability and efficiency. Table 6.1 summarises the T-type MLI's switching method.

Table 6.1 Switching states of T-type MLI.

Voltage level	Conducting switches
V_{dc}	T_1, T_2
V_{dc} /2	T_5, T_6, T_2
0	T_1, T_3 / T_2, T_4
$-V_{dc}$ /2	T_3, T_6, T_5
$-V_{dc}$	T_3, T_4

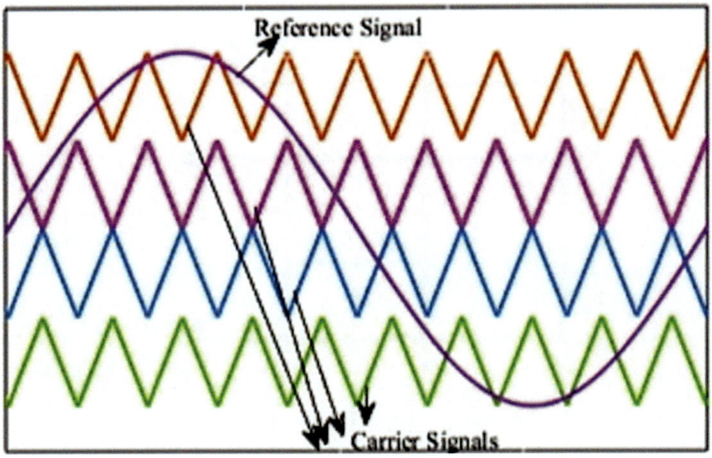

Figure 6.4 Generation of switching signals using MC-SPWM.

Table 6.2 Switching paths of T-type MLI.

Voltage level	Current paths
V_{dc}	$C_{1+} - T_1 -$ Load $- T_2 \quad - C_2 - C_{1-}$
$V_{dc}/2$	$C_{2+} - T_5 - T_6 -$ Load $- T_2 - C_{2-}$
0	*Current circulates* $T_1 - T_3$ / $T_2 - T_4$
$-V_{dc}/2$	$C_{1+} - T_3 -$ Load $- T_6 - T_5 - C_{1-}$
$-V_{dc}$	$C_{1+} - T_3, -$ Load$- T_4 - C_2 - C_{2-}$

Multi-carrier sinusoidal PWM (MC-SPWM) has been employed to reduce power losses in contrast to unipolar PWM [14], and the associated switching signals are illustrated in Figure 6.4. The current paths for the generation of different levels are tabulated in Table 6.2.

6.3 Simulation Results

The operation of the dual input topology is validated with simulation performed in MATLAB/Simulink environment with the parameters listed in Table 6.3.

For an exhaustive analysis, various scenarios have been considered. To begin, a change in irradiance is taken into account on one of the panels. When both panels were initially exposed to the same irradiance levels (800 W/m^2), both panels extracted a maximum power of 118 W each, and the panels shared the load equally. The irradiance shift on the second panel is taken into account at $t = 0.7$ sec. It is found that the first panel extracts 116.8 W

Table 6.3 Parameters used.

Parameter	Value	Parameter	Value
V_{mpp} (V)	34.5	I_{mpp} (A)	4.35
P_{mpp} (W)	150	L_{dc} (mH)	2.5
C_{pv1} and C_{pv2} (μF)	100	C_1 and C_2 (μF)	1200
Load (Ω)	10	f_s (DC stage) (kHz)	10
f_c (Carrier frequency) (kHz)	5	C_{par1} and C_{par2} (nF)	10

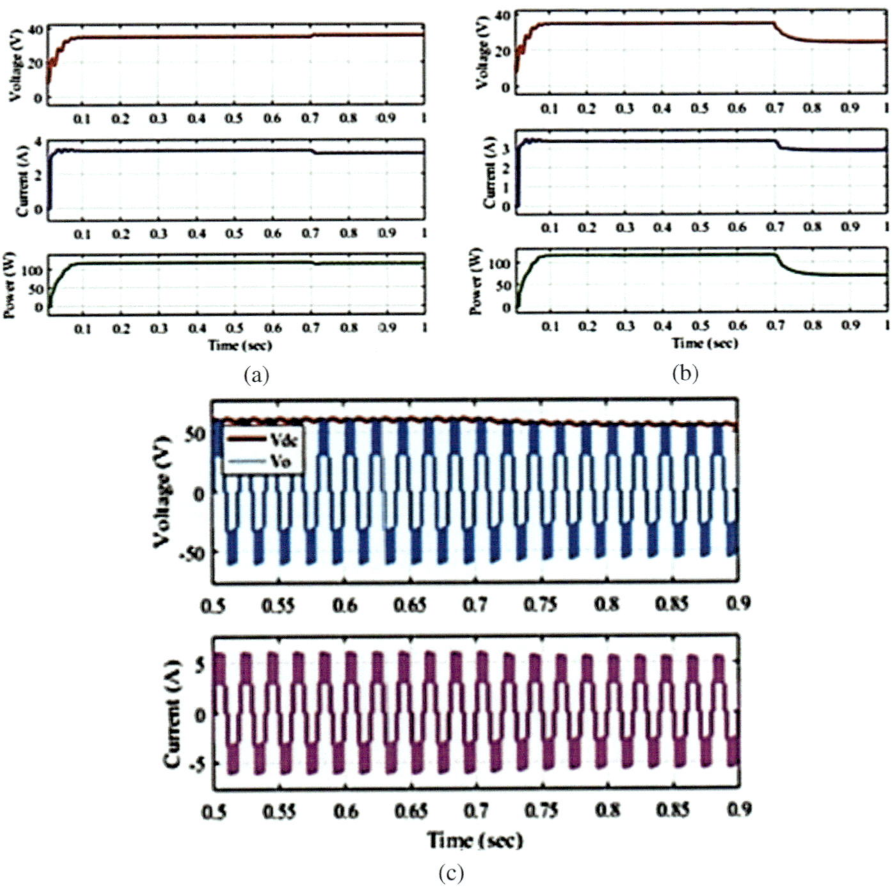

Figure 6.5 Parameters of (a) panel 1, (b) panel 2, and (c) DC link and load.

near its MPP, while the second panel extracts 88.2 W near its MPP due to irradiance change (600 W/m^2). In partial shading situations, the proposed dual input arrangement is revealed to draw the maximum power from each panel. The reduction in irradiance by 25% led in a V_0 and I_0 reduction of

18% and 7.2%, respectively. Figure 6.5 depicts this scenario. Panel 1 and 2's voltage, current, and power are depicted in Figure 6.5 (a) and Figure 6.5 (b), respectively. Variations in irradiance cause changes in panel voltage and, as a result, the voltage varies at the DC link. As demonstrated in Figure 6.5 (c), the fluctuation in DC-link voltage reflects the load voltage and current.

Second, dynamic loading is taken into account. As the load increases, higher currents are required, resulting in a drop in the DC-link voltage and, as a result, the load voltage, as illustrated in Figure 6.6. The change in load of 18% at $t = 0.7$ sec resulted in a 16.6% increase in current and a 3.7% reduction in DC link and load voltage, as shown in Figure 6.6. It can be seen that the panels deliver equal amounts of power to the load before and after the load change, demonstrating the inherent capability of power sharing.

Various approaches for lowering leakage currents below VDE 0126–1–1 German norms (300 mA) have been presented in the literature [15].

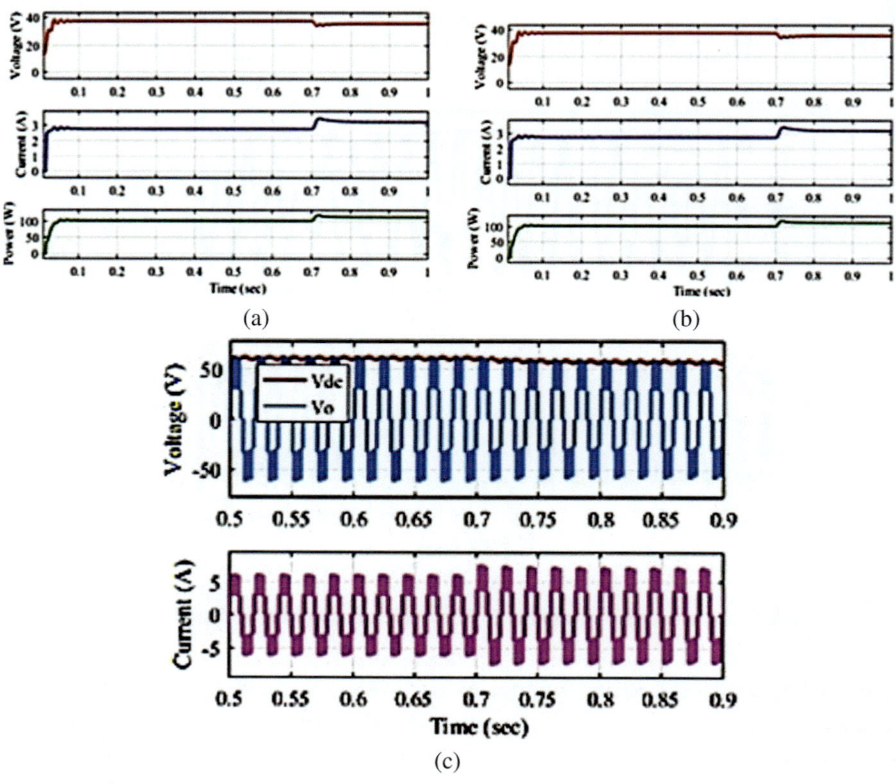

Figure 6.6 Parameters of (a) panel 1, (b) panel 2, and (c) DC link and load.

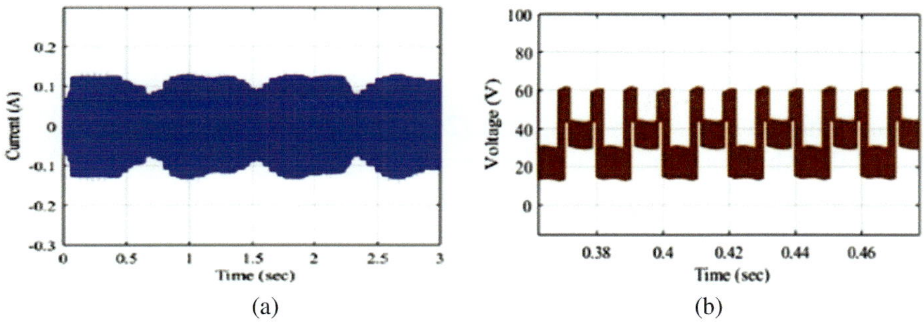

Figure 6.7 Waveform of (a) leakage current and (b) common mode voltage.

The proposed structure has the benefit that the leakage currents are less than 300 mA, and the modulation approach has further decreased them, as depicted in Figure 6.7 (a). The inverter's common mode voltage is shown in Figure 6.7 (b).

The simulation verifies the extraction of maximum power through each panel for different operating conditions, as well as the leakage current being within German norms.

6.4 Control Strategy

The proposed system employs two control techniques: MPPT control, which aims to achieve MPPT of individual modules, and inverter control, which focuses on regulating grid voltage.

6.4.1 MPPT control

The system utilizes a time-multiplexing switching technique to ensure individual module operating at its MPP during partial shading, thereby preventing series operation of the PV modules. The traditional perturb and observe (P&O) method is implemented for tracking maximum power because it has advantages such as flexibility, ease of implementation, and low computational effort [16].

6.4.2 Inverter control

On the AC side, a pre-defined voltage is required for efficient load functioning. In the contemporary, for regulation of load voltage, a variety of inverter control approaches have been reported. A few relate to control of dead beat [17], repetitive control [18], and sliding mode control [19] that have become widely popular because of their superior dynamic performance. They also provide a quicker response and, if properly designed, can reduce overshoot and ringing. These, on the other hand, are extremely sensitive, complicated, and cannot mitigate the steady-state error [20–21]. Increased dynamic performance while exhibiting null steady-state error is achieved by the synchronous reference frame (SRF) controller [22–23]. In three-phase systems, the SRF-based PI controller is reported in a lot of literature; however, there exists a comparatively scarce body of literature concerning the utilization of the SRF-PI methodology in single-phase systems. This technology is utilized for suggested T-type inverter which is connected to grid for attaining voltage regulation, thanks to the advantage of mitigating steady-state error. For grid frequency management, a PLL is used and regulated.

6.4.2.1 Generation of orthogonal phase

Park's transformation is used to convert three signals into d-q coordinates in three-phase systems. For single-phase systems, however, an orthogonal phase is required to generate the d-q frame. Several ways, such as enabling the original signal to be delayed [24], differentiating the original signal [25], second-order generalized transform integrator [26], all pass method [27], Hilbert transform [28], and Kalman filter method [26] were presented for the generation of orthogonal signals. A delay is introduced to the original signal and verified for the proposed controller, aiming to simplify its implementation. As a result, the original signal is combined with the created orthogonal signal to convert from $\alpha\beta$- to dq-frame.

6.4.2.2 Control strategy

Figure 6.8 depicts the control structure that is employed to regulate the output voltage. For immediate output voltage regulation, the approach employs an SRF-PI-based controller. For dampening and better transient responsiveness, an inner loop is included to control the current. A feed-forward path for voltage, in addition to the two loops, ensures the system's robustness against parameter changes. To activate the inverter over current protection, the inner current loop must be implemented.

6.4.2.3 Controller design

Commencing with the generation of an orthogonal signal, the controller design specifics are outlined as follows:

Step 1: The available voltage signal (V_α) is used to generate the orthogonal signal (V_β) as

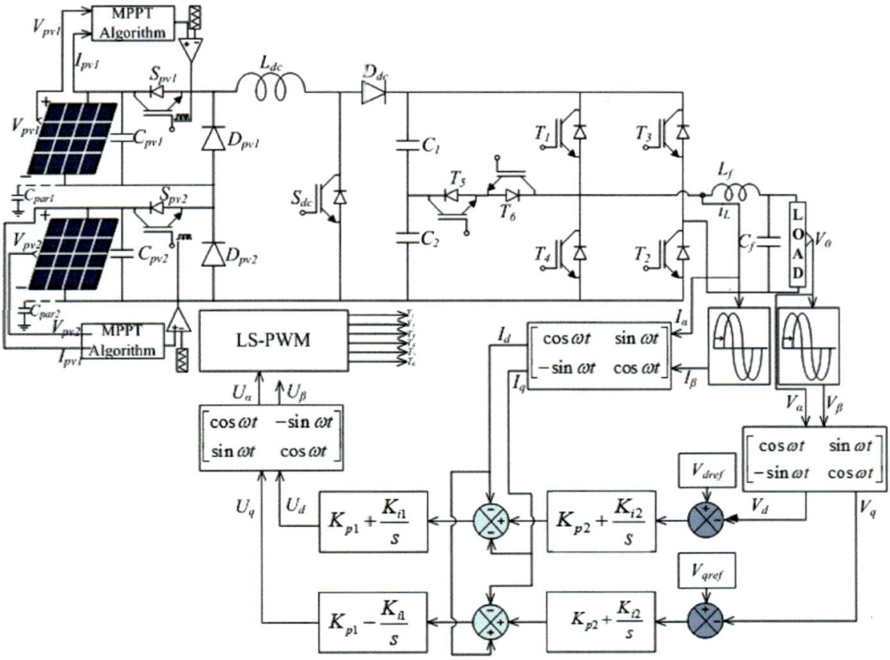

Figure 6.8 Proposed structure with control strategy.

$$V_\alpha = V_o \sin \omega t, \tag{6.2}$$

$$V_\beta = V_o \sin (\omega t - 90) . \tag{6.3}$$

Step 2: Park's transformation is used to translate the synchronous reference ($\alpha\beta$) to a stationary reference (dq) frame. Utilizing a standard DC regulator like a conventional PI, the synchronous reference frame switches fundamental frequency data to DC, thereby mitigating steady-state error. The transformation of the Park in a single phase is given by

$$\begin{bmatrix} V_d \\ V_q \end{bmatrix} = \begin{bmatrix} \cos \omega t & \sin \omega t \\ -\sin \omega t & \cos \omega t \end{bmatrix} \begin{bmatrix} V_\alpha \\ V_\beta \end{bmatrix} . \tag{6.4}$$

Step 3: The design of the controller for the α-frame is elaborated upon, with similar principles employed for β-frame controller design. The inner loop's transfer function is:

$$G_i(s) = \left(\frac{\frac{1}{L_f}}{s + \frac{r_{L_f}}{L_f}} \right) \left(K_{p1} + \frac{K_{i1}}{s} \right) . \tag{6.5}$$

At the gain cross-over frequency, the magnitude of the transfer function is

$$|G_i(j\omega)| = 1, \tag{6.6}$$

$$\left| \frac{\frac{1}{L_f}}{s + \frac{r_{L_f}}{L_f}} \right| \left| K_{p1} + \frac{K_{i1}}{s} \right| \Bigg|_{s=j\omega_{\text{BWI}}} = 1. \tag{6.7}$$

Here, ω_{BWI} represents the bandwidth of the inner loop, typically, selected as:

$$\omega_{\text{BWI}} = \frac{1}{10} \times 2pf_s \text{ rad/sec}, \tag{6.8}$$

where, f_s is the switching frequency. Substituting the converter parameters in eqn (6.7), the objective function formulated for the optimization problem is:

$$min(\ f_1(K_{p1}, K_{i1})) = K_{p1} + (3.18 \times 10^{-4})K_{i1} - 3.92. \tag{6.9}$$

The genetic algorithm (GA), a well-utilized evolutionary optimization tool, is employed to minimize f_1. The *[K_{p1}, K_{i1}]* solution acquired after GA convergence is used as the parameters for current controller.

Step 4: Increasing ω_{BWI} enables a swifter reaction to potential current variations and permits the adjustment of outer loop (voltage) parameters based on the unity transfer function of the entire inner loop. The objective function is stated as eqn (6.11), and the transfer function is derived (eqn (6.10)) in the same way as in step 3.

$$\left| \frac{1}{Cs} \right| \left| K_{p2} + \frac{K_{i2}}{s} \right| \Bigg|_{s=j\omega_{\text{BWI}}} = 1, \tag{6.10}$$

where ω_{BWO} is the bandwidth of the outer voltage loop, $\omega_{\text{BWO}} \ll \omega_{\text{BWI}}$, generally $\omega_{\text{BWO}} < \frac{1}{3}\omega_{\text{BWI}}$

$$min(\ f_2(K_{p2}, K_{i2})) = K_{p2} + (2.5 \times 10^{-4})K_{i2} - 0.8. \tag{6.11}$$

Step 5: The indices in dq-frame are transformed to $\alpha\beta$-frame as

$$\begin{bmatrix} V_\alpha \\ V_\beta \end{bmatrix} = \begin{bmatrix} \cos\omega t & -\sin\omega t \\ \sin\omega t & \cos\omega t \end{bmatrix} \begin{bmatrix} V_d \\ V_q \end{bmatrix}. \tag{6.12}$$

Employing the LS-PWM switching methodology, the inverter gate pulses are produced when control output is combined with carrier waves.

6.5 Simulation Results

Simulation model developed in MATLAB/Simulink is used to validate the efficacy of the proposed topology as well as the control strategy. For the purpose of validating the suggested technique for load and source variations, two instances were investigated. The converter parameters are presented in Table 6.3, whereas the PV module specifications are listed in Table 6.4.

Table 6.4 Parameters of the module.

Indices	Value	Indices	Value
T	38 °C	P_{mpp}	150 W
I_{sc}	5.4 A	V_{mpp}	34.5 V
I_{mpp}	3.5 A	V_{oc}	42.5 V

6.5.1 Case 1: Partial shading condition on one of the modules

The previous section has been considered to validate the effective operation of the controller. Initially, the PV modules were operated at an irradiance of 821 W/m^2, with one module's irradiance reduced to 621 W/m^2 to accommodate partial shading. As shown in Figure 6.9, the suggested integrated structure efficiently accomplishes the MPPT of individual modules at both irradiance levels. Module 1 and module 2 responses are illustrated in Figure 6.9 (a) and Figure 6.9 (b), respectively. Both modules start out from 120 W of power, but after partial shade on module 1, the power extracted on module 1 is 88 W and module 2 is 118 W, indicating that each module's operation is closer to individual MPP as discussed in previous section. This demonstrates that

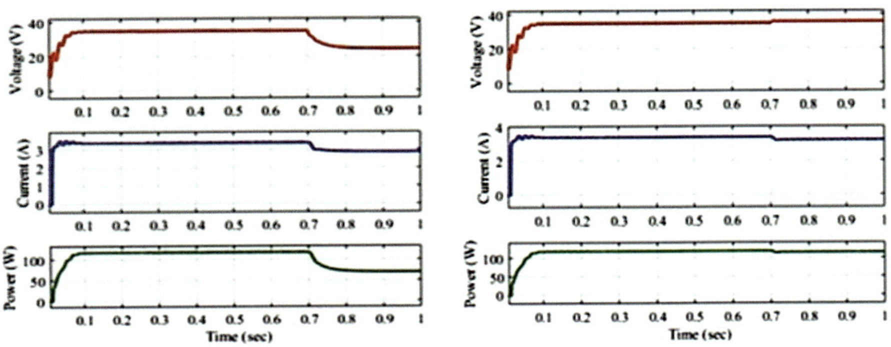

(a) (b)

Figure 6.9 Parameters of (a) module 1 and (b) module 2.

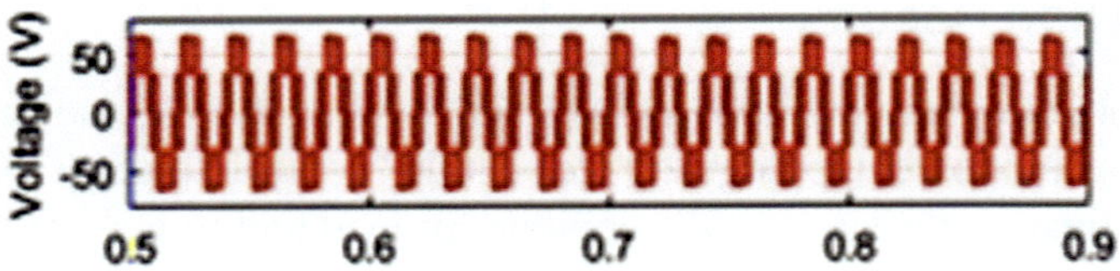

(a)

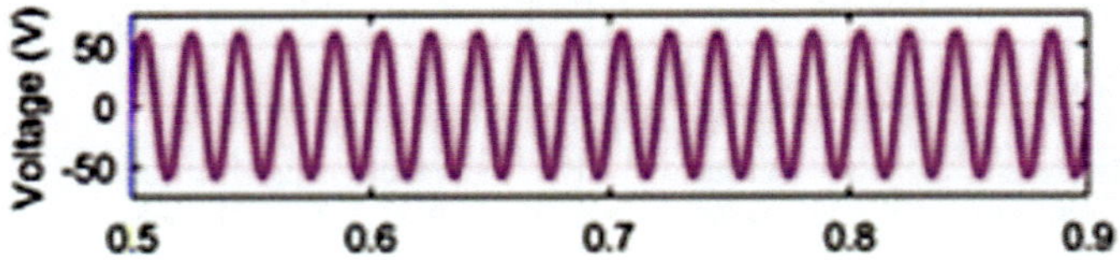

(b)

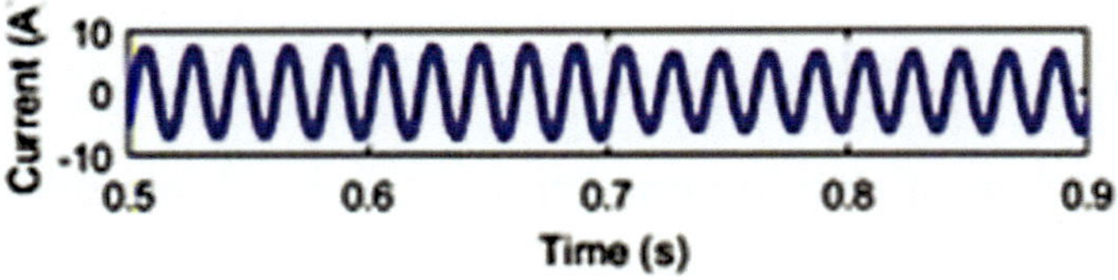

(c)

Figure 6.10 (a) Output voltage of MLI. (b) Grid voltage. (c) Current to the grid.

the proposed topology achieves MPP of individual modules under partial shading. Figure 6.10 (a) shows the output voltage of a T-type interfaced inverter. As illustrated in Figure 6.10 (b), the voltage response utilizing the developed controller is regulated at the desired value regardless of partial shade on the PV module. The current, on the other hand, decreases as a result of the lower generated power, as seen in Figure 6.10 (c).

6.5.2 Case 2: Dynamic loading scenarios

To assess the performance of controller for dynamic loading scenarios 80 W load is added to existing 200 W load at 0.6 s and removed at 0.8 s. Figure 6.11 shows the voltage and current response to a load fluctuation circumstance. As the load rises, the current also increases, while ensuring a stable five-level voltage, illustrated in Figure 6.11 (a). Regardless of the change in power demand, the controller maintains the specified load voltage, as shown in Figure 6.11 (b). Figure 6.11 (c) depicts the change in the current provided to the grid.

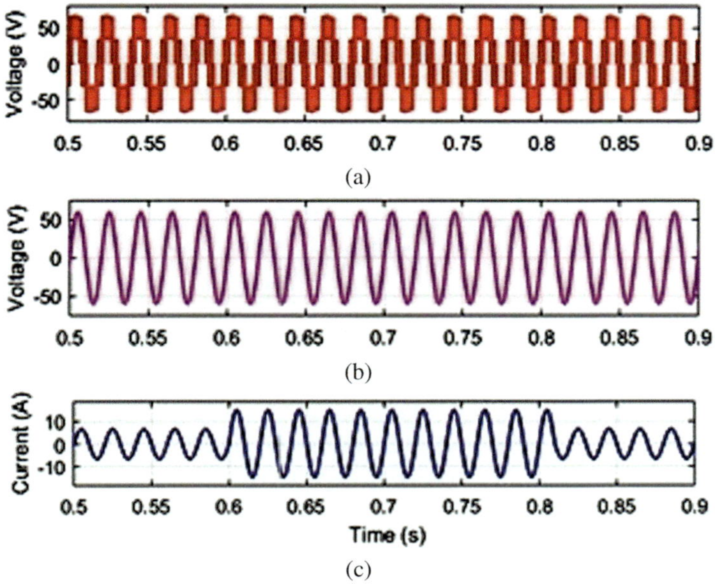

Figure 6.11 (a) Output voltage of MLI. (b) Grid voltage. (c) Current to the grid.

6.6 MPPT Efficiency Analysis

The power supplied by every individual unit highly depends on external circumstances and the MPP at various levels of irradiance levels that is given by [29]

$$P_{\text{expected}} = P_m \left(\frac{G}{G_o} \right), \tag{6.13}$$

where G_o is the standard test irradiance at 1000 W/m^2, P_m denotes the rated maximum power of module and G is the irradiance at the instant. The MPP efficiency (tracking precision) is given by [30]

$$\eta_{\text{MPP}} = \frac{P_{PV}}{P_{\text{mpp}}} \times 100. \tag{6.14}$$

The theoretical and experimental power extracted for various irradiance levels is presented in Figure 6.12. It is worth noting that, both the MPPs closely match and validates the effectiveness of proposed scheme for various levels of irradiance.

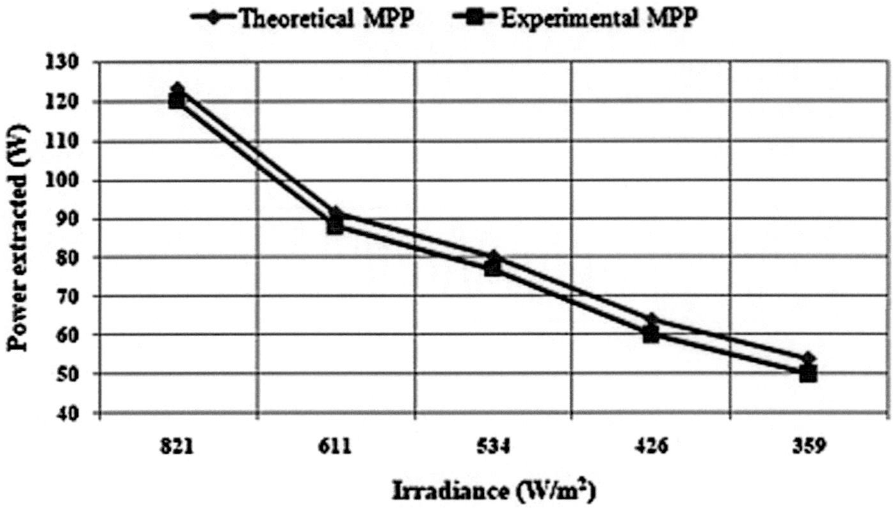

Figure 6.12 MPPT efficiency graph.

6.7 Conclusion

This chapter discussed a dual input-fed transformerless multilevel inverter for standalone PV applications. Under partial shading conditions, the suggested dual input structure efficiently harvests the maximum power. The proposed multi-input structure has the capacity to distribute load evenly. A T-type inverter with voltage-dividing capacitors as DC link capacitor was employed to obtain five levels in the output voltage. The suggested system has been tested in MATLAB/Simulink for various operational conditions. The proposed dual input structure can be extended to many inputs, and effective switching strategy selection helps to extract maximum power in various operating conditions. Further, design of controller so as to regulate the output voltage at the desired value.

References

[1] J. I. Leon and D. Vinnikov. (2015) Grid-Connected Photovoltaic Systems: An Overview of Recent Research and Emerging PV Converter Technology IEEE Ind. Electron. Mag., no. March, pp. 47–61, doi: 10.1109/MIE.2014.2376976.

[2] W. Li, Y. Gu, H. Luo, W. Cui, and X. He. (2015) Topology Review and Derivation Methodology of Single-Phase Transformerless Photovoltaic

Inverters for Leakage Current Suppression IEEE Trans. Ind. Electron., vol. 62, no. 7, pp. 4537–4551, doi:10.1109/TIE.2015.2399278.

[3] B. Yang, W. Li, Y. Zhao, and X. He. (2010) Design and Analysis of a Grid-Connected Photovoltaic Power System IEEE Trans. Power Electron., vol. 25, no. 4, pp. 992–1000.

[4] N. Vazquez, J. Vazquez, J. Vaquero, C. Henrandez, and E. Vazquez. (2017) Integrating Two Stages as a Common-Mode Transformerless Photovoltaic Converter IEEE Trans. Ind. Electron., vol. 64, no. 9, pp. 7498–7507.

[5] G. R. Walker and P. C. Sernia. (2004) Cascaded DC – DC Converter Connection of Photovoltaic Modules IEEE Trans. Power Electron., vol. 19, no. 4, pp. 1130–1139.

[6] S. M. Macalpine, R. W. Erickson, and M. J. Brandemuehl. (2013) Characterization of Power Optimizer Potential to Increase Energy Capture in Photovoltaic Systems Operating Under Nonuniform Conditions IEEE Trans. Power Electron., vol. 28, no. 6, pp. 2936–2945, doi:10.1109/TPEL.2012.2226476.

[7] B. Xiao et al. (2015) Modular Cascaded H-Bridge Multilevel PV Inverter With Distributed MPPT for Grid-Connected Applications IEEE Trans. Ind. Appl., vol. 51, no. 2, pp. 1722–1731, doi:10.1109/TIA.2014.2354396.

[8] Q. Huang, W. Yu, and A. Q. Huang. (2016) "Independent DC Link Voltage Control of Cascaded Multilevel PV Inverter," 2016 IEEE Appl. Power Electron. Conf. Expo., pp. 2727–2734, doi:10.1109/APEC.2016.7468249.

[9] B. Liu, S. Duan, and T. Cai. (2011) Photovoltaic DC-Building-Module-Based BIPV System — Concept and Design Considerations IEEE Trans. Power Electron., vol. 26, no. 5, pp. 1418–1429, doi:10.1109/TPEL.2010.2085087.

[10] W. Xiao, N. Ozog, and W. G. Dunford. (2007) Topology Study of Photovoltaic Interface for Maximum Power Point Tracking IEEE Trans. Ind. Electron., vol. 54, no. 3, pp. 1696–1704

[11] H. Hu, S. Harb, S. Member, N. Kutkut, I. Batarseh, and Z. J. Shen. (2013) A Review of Power Decoupling Techniques for Microinverters With Three Different Decoupling Capacitor Locations in PV Systems IEEE Trans. Power Electron., vol. 28, no. 6, pp. 2711–2726, doi:10.1109/TPEL.2012.2221482.

[12] M. A. Rezaei, K. Lee, and A. Q. Huang. (2016) A High-Efficiency Flyback Micro-inverter With a New Adaptive Snubber for Photovoltaic

Applications IEEE Trans. Power Electron., vol. 31, no. 1, pp. 318–327, doi:10.1109/TPEL.2015.2407405.

[13] L. Kumar and S. Jain. (2013) A multiple source DC/DC converter topology Int. J. Electr. Power Energy Syst., vol. 51, pp. 278–291, doi: 10.1016/j.ijepes.2013.02.020.

[14] J. Lee, R. Kwak, and K. Lee. (2018) Novel Discontinuous PWM Method for a Single-Phase Three-Level Neutral Point Clamped Inverter With Efficiency Improvement and Harmonic Reduction IEEE Trans. Power Electron., vol. 33, no. 11, pp. 9253–9266, doi:10.1109/TPEL.2018.27 94547.

[15] G. Petrone et al. (2008) Reliability Issues in Photovoltaic Power Processing Systems IEEE Trans. Ind. Electron., vol. 55, no. 7, pp. 2569–2580, doi:10.1109/TIE.2008.924016.

[16] M. Khaled, H. Ali, M. Abd-El Sattar, and A. A. Elbaset. (2016) Implementation of a modified perturb and observe maximum power point tracking algorithm for photovoltaic system using an embedded microcontroller IET Renew. Power Gener., vol. 10, no. 4, pp. 551–560, doi:10.1049/iet-rpg.2015.0309.

[17] A. Kawamura and K. Ishihara. (1988) "Real time digital feedback control of three phase PWM inverter with quick transient response suitable for uninterruptible power supply," Conf. Rec. 1988 IEEE Ind. Appl. Soc. Annu. Meet., pp. 728–734, doi:10.1109/IAS.1988.25143.

[18] R. Ortega, G. Garcera, E. Figueres, O. Carranza, and C. L. Trujillo. (2011) "Design and application of a two degrees of freedom control with a repetitive controller in a single phase inverter," Proc. - ISIE 2011 2011 IEEE Int. Symp. Ind. Electron., pp. 1441–1446, doi:10.1109/ISIE .2011.5984372.

[19] T. L. Tai and J. S. Chen. (2002) UPS inverter design using discrete-time sliding-mode control scheme," IEEE Trans. Ind. Electron., vol. 49, no. 1, pp. 67–75, doi:10.1109/41.982250.

[20] K. Zhou and Yang Yunhu "Phase Compensation Multiple Resonant Control of Single-phase PWM Inverter," 2012, pp. 453–457.

[21] S. A. Khajehoddin, M. Karimi-Ghartemani, P. K. Jain, and A. Bakhshai. (2012) A resonant controller with high structural robustness for fixed-point digital implementations IEEE Trans. Power Electron., vol. 27, no. 7, pp. 3352–3362, doi:10.1109/TPEL.2011.2181422.

[22] M. J. Ryan and R. D. Lorenz. (1997) "Synchronous-frame controller for a single-phase sine wave inverter," Conf. Proc. - IEEE Appl. Power

Electron. Conf. Expo. - APEC, vol. 2, pp. 813–819, doi:10.1109/APEC.1997.575739.

[23] M. Monfared, S. Golestan, and J. M. Guerrero. (2014) Analysis, design, and experimental verification of a synchronous reference frame voltage control for single-phase inverters IEEE Trans. Ind. Electron., vol. 61, no. 1, pp. 258–269, doi:10.1109/TIE.2013.2238878.

[24] R. Zhang, M. Cardinal, P. Szczesny, and M. Dame. (2002) "A Grid Simulator with control of single-phase power converters in D-Q rotating frame," PESC Rec. Annu. Power Electron. Spec. Conf., vol. 3, pp. 1431–1437, doi:10.1109/PSEC.2002.1022377.

[25] A. Roshan, R. Burgos, A. C. Baisden, F. Wang, and D. Boroyevich. (2007) "A D-Q frame controller for a full-bridge single phase inverter used in small distributed power generation systems," Conf. Proc. - IEEE Appl. Power Electron. Conf. Expo. - APEC, pp. 641–647, doi: 10.1109/APEX.2007.357582.

[26] M. Ciobotaru, R. Teodorescu, and F. Blaabjerg. (2006) "A new single-phase PLL structure based on second order generalized integrator," PESC Rec. - IEEE Annu. Power Electron. Spec. Conf., pp. 1–6, doi: 10.1109/PESC.2006.1711988.

[27] Rae-Young Kim, See-Young Choi, and In-Young Suh. (2004) "Instantaneous control of average power for grid tie inverter using single phase D-Q rotating frame with all pass filter," 30th Annu. Conf. IEEE Ind. Electron. Soc. 2004. IECON 2004, vol. 1, pp. 274–279, doi:10.1109/IECON.2004.1433322.

[28] M. Saitou and T. Shimizu. (2002) "Generalized theory of instantaneous active and reactive powers in single-phase circuits based on Hilbert transform," 2002 IEEE 33rd Annu. IEEE Power Electron. Spec. Conf. Proc., no. 2, pp. 1419–1424, doi:10.1109/PSEC.2002.1022375.

[29] R. Hariharan, M. Chakkarapani, G. Saravana Ilango, and C. Nagamani. (2016) A Method to Detect Photovoltaic Array Faults and Partial Shading in PV Systems IEEE J. Photovoltaics, vol. 6, no. 5, pp. 1278–1285, doi:10.1109/JPHOTOV.2016.2581478.

[30] B. Subudhi and R. Pradhan. (2013) A Comparative Study on Maximum Power Point Tracking Techniques for Photovoltaic Power Systems Sustain. Energy, IEEE Trans., vol. 4, no. 1, pp. 89–98, doi:10.1109/TSTE.2012.2202294.

7

Smart Grid Communication and Protocols and Smart Interfacing of Electric Vehicles with the Grid

Kushan Tharuka, Ruvini De Seram, Anindita Golder, and Sheldon Williamson

Smart Transportation Electrification and Energy Research(STEER) Group
Department of Electrical, Computer, and Software Engineering Ontario Tech University Oshawa, Canada
E-mail: kushan.lulbaddawaduge@ontariotechu.net;
edirisooriyamohottige.deseram@ontariotechu.net;
anindita.golder@ontariotechu.net; sheldon.williamson@ontariotechu.ca

Abstract

The development of smart grids and smart cities have been aided by the rapid development of internet-of-things (IoT), wireless sensors and fog computing technologies. These technologies have allowed us to integrate advanced metering infrastructure (AMI) and develop smart energy management systems (EMS) with the existing power systems for real time operation of microgrids. The power flows are further enabled by the power electronic converters that enable the integration of several distributed energy resources (DER) such as renewable energy systems (RES) and electric vehicles (EV) with the grid. As such, this chapter aims to explain the technological advancements in the areas of AMI, EMS and power converters for grid integration of microgrids.

Keywords: Smart grids, smart meters, advanced metering, power converters, energy management systems, electric vehicles, internet-of-things.

7.1 Introduction

The rapid penetration of renewable energy systems (RES) and electric vehicles (EV) has led to the development of microgrids and nanogrids. This has led to the transition from a conventional power system where the power flow was unidirectional to a modernized power system with bidirectional power flows as consumers can now become producers (known as prosumers). As shown in Figure 7.1 and according to the International Energy Agency (IEA) [1], "smart grids are electricity networks that use digital technologies, sensors, and software to better match the supply and demand of electricity in real-time while minimizing costs and maintaining the stability and reliability of the grid."

The transition toward an integrated and smart grid system has been aided by the development of technologies such as Internet of things (IoT), communication systems, blockchain technology, cloud computing, fog computing, machine learning and artificial intelligence, and power electronic converters and control systems. The applications of these technologies include smart meters and advanced metering infrastructure (AMI), energy management systems (EMS), electricity markets, and integrating EVs with the grid. Accordingly, this chapter focuses on the developments of the abovementioned technologies for converting the traditional grid to a smart grid.

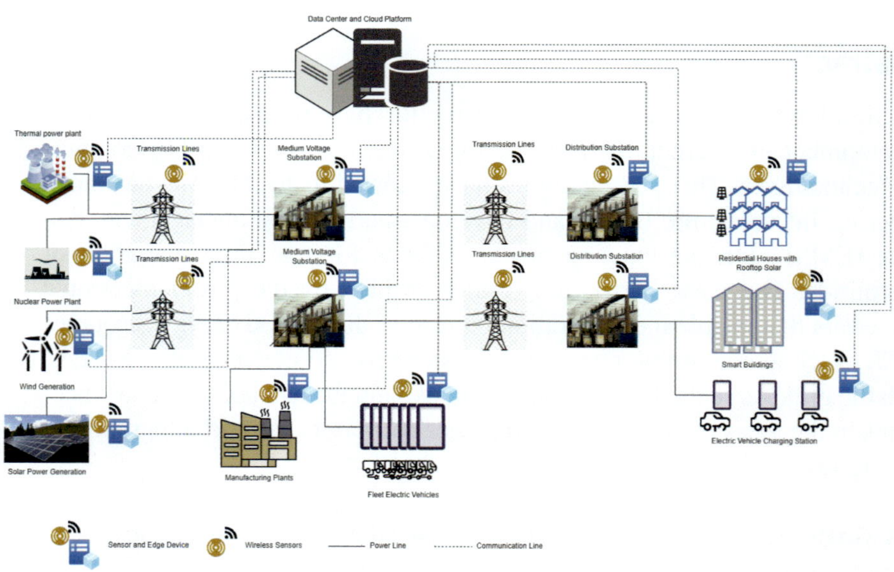

Figure 7.1 Smart grid architecture.

The development of smart grids would not be possible without the integration of smart meters, wireless sensors, edge and cloud computing devices, and AMI. However, the integration of these devices comes with its own set of challenges such as data privacy, cyber threats, and so on. Hence it is necessary to look at the advancements in these technologies.

7.2 Advanced Metering Infrastructure (AMI)

Growth in economies and population drives up power demand. Smart grid technologies make the grid more responsive and efficient, which helps utilities better manage the increasing electricity demand. However, many obstacles confront the electrical industry, including increased energy consumption, high aggregate technical and commercial (AT&C) losses, concerns about the quality and dependability of the power supply, fuel restrictions, and the requirement to follow environmental regulations aimed at halting climate change. These challenges have prompted a shift in recognizing both consumers and utility providers as intelligent energy decision-makers, with a focus on advancing real-time energy efficiency. Smart grid technology gives utility companies and customers access to real-time data and decision-making tools. Customers may use this information to make well-informed decisions about how much energy they use, and utilities can use it to streamline operations, forecast demand, and put energy-saving measures into place.

The smart grid design works as a closed-loop control system, with data produced by tracking loading patterns on the distribution side serving as feedback to the grid controller and electricity provided on the demand side serving as the system output. This closed-loop power system model can be seen as a hierarchical network architecture with the master data management system (MDMS) for big data management, neighborhood area networks (NANs) for load management and data collection, and wide area networks (WANs) for data processing and storage, as illustrated in Figure 7.2. AMI, or advanced metering infrastructure, is one of the fundamental components of a smart grid. It can be defined as a system designed to measure, collect, transfer, and analyze energy usage data. It involves communication with metering devices, allowing for real-time or near-real-time exchange of information. The role of AMI in the smart grid is to provide advanced data collecting and transmission capabilities, which makes the electrical grid more responsive and intelligent.

Measurement of energy usage, data collection from metering devices, data transfer to utility networks, and data analysis are among the vital tasks

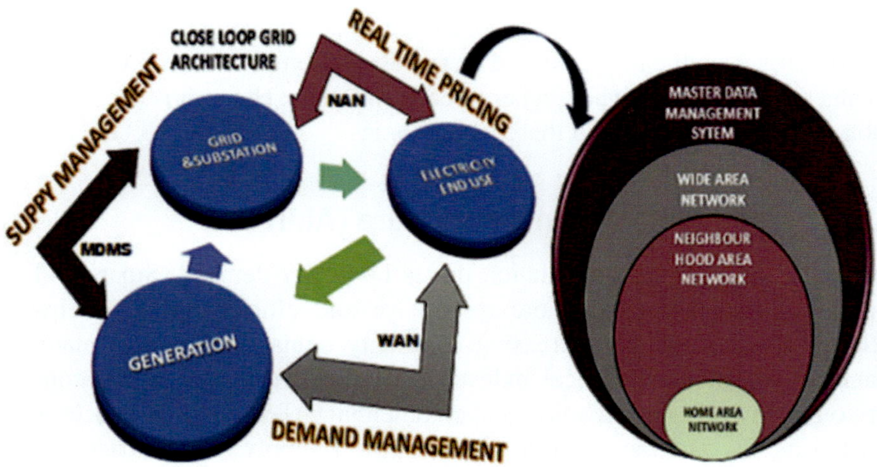

Figure 7.2 Closed-loop smart grid network architecture.

carried out by AMI. Utilities are therefore equipped to decide on grid operations and energy management with knowledge. AMI's capacity to involve end users in the energy management process is one of its main benefits. AMI gives customers the capacity to actively contribute to lowering peak demands and make educated decisions about their energy consumption by giving them access to real-time information on energy usage. Furthermore, utilities benefit from improved grid status visibility using AMI. They can keep an eye on trends in energy usage, spot high-demand locations, and react quickly to any problems or disruptions. AMI systems' data collection and analysis empower utilities to make data-driven choices about overall energy-saving initiatives, infrastructure development, and grid operations.

7.2.1 Key components of AMI infrastructure

The key components that make up an AMI infrastructure include the smart meter, communication channel, meter data acquisition system/meter data management (MDAS/MDM), load control, demand response, load monitoring, tamper detection, alarm handling, real-time energy audit, Time of day (ToD) tariff and so on. Figure 7.3 illustrates that smart meters are the foundational devices in AMI, which can measure and gather comprehensive data on energy use. They facilitate information sharing between the utility and end consumers by offering two-way communication. The transmission

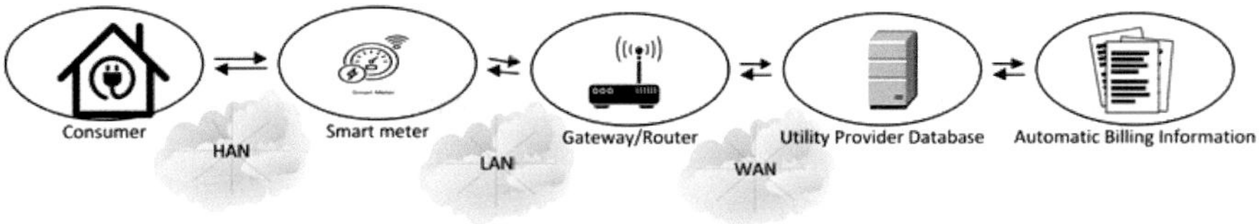

Figure 7.3 Basic AMI system structure showing smart metering process.

of data between utility systems and smart meters requires the communication medium. This can involve a variety of communication technologies, including radio frequency, cellular networks, powerline communication, and combinations of these. Metering data is collected, stored, and processed by MDAS/MDM systems. They are essential in organizing and providing utility operations and analysis with access to the massive volumes of data generated by smart meters.

Load monitoring involves tracking and analyzing the energy consumption patterns of users. This capability provides valuable insights into peak demand periods, allowing utilities to optimize grid operations and plan for capacity upgrades. During times of peak demand, utilities may actively monitor and regulate energy use thanks to demand response capabilities. This may involve offering incentives to users to use less power during periods of high demand. In order to balance the load on the grid, utilities can remotely monitor and operate certain devices, such as HVAC systems or appliances. This enhances the efficiency and stability of the grid. Features for detecting tampering with smart meters aid in detecting unwanted access or interference. To keep people confident in the system, metering data security and integrity must be guaranteed. Alarm handling systems notify utilities of anomalous circumstances or possible grid problems. Rapid reaction to alerts enhances the power distribution system's overall resilience and dependability.

7.2.1.1 Smart metering

The smart meter is an electronic device that tracks and measures a customer's power usage, sending data to the utility company for billing purposes. Smart metering is thought to be a catalyst for digitalization adoption and energy efficiency. Between utility companies and customers, smart energy meters serve as an essential conduit for communication. In the energy distribution system, two-way communication makes interactions more dynamic and responsive. A meter's main function is to monitor parameters and energy; however, smart meters also produce a significant quantity of data. Making better-educated decisions is made possible by the comprehensive insights into the complete electricity delivery system that this high-resolution data offers. Instead of using electromechanical technology, smart electricity meters employ digital technology. Modern measuring or metering equipment and smart meters vary primarily in that the former assumes control over operations.

Installed at the sites of the customers, smart meters function as data sources that offer real-time data on energy use, utility-based power consumption, power quality, and tariff specifics. Via a neighborhood area network (NAN), the local distribution company's (LDC) server receives the data gathered from smart meters. This guarantees that information is shared between customers and the utility in almost real-time. The gathered data is then forwarded to the grid's service provider for in-depth analysis and kept in meter data management/repository (MDM/R) systems. This makes it possible to thoroughly analyze the data and use it for a variety of goals. The amount of data being produced by these devices is shown by the fact that 46 million smart meters are gathering one billion data points daily in the United States [2]. Certain nations, such as Germany, have implemented smart meter technology and imposed legal obligations on users who utilize energy beyond 6000 kWh [3]. The AMI system has potential and difficulties as a result of this enormous information. The amount of data is rising, and it is clear that the current AMI infrastructure has to be continuously improved in order to handle and manage the growing amount of data from smart meters.

Smart meters have the ability to provide the utility readings of a customer's actual power use. Securing this communication link from attacks remains very important for the secure transmission of readings. Many techniques and methodologies have been developed, with large computational overheads in the majority of the research. Authors have proposed a lightweight concealed based security scheme (CBSS) for secure transmission inside the AMI in order to provide authentication, minimize computational overheads, and use less energy during transmission [3]. The proposed network model for AMI infrastructure is shown in Figure 7.4.

The sensors, base stations (BSs), web servers (WSs), application providers (APs), Internet portals (IPs), and users linked to smart meters (SMs) comprise the hierarchical interconnections of the AMI system. Data on consumption is measured by the smart meter's sensors and transmitted to the BS over a secure connection. The relevant users, who are the clients, are subsequently given access to the data. There is now a multi-hop connection between the users and the sensors. As seen in Figure 7.3, an Internet gateway links the BS and the WS in a hierarchical arrangement. The utility provider

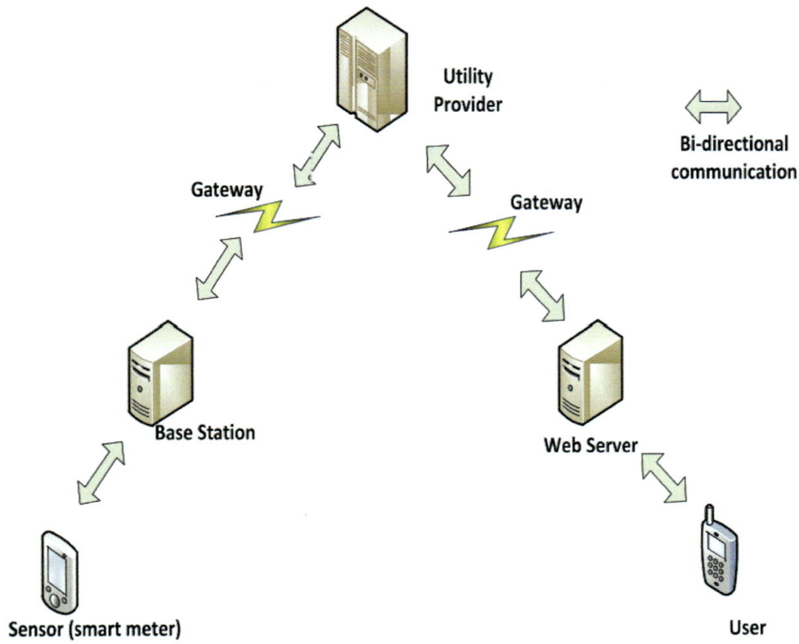

Figure 7.4 Network architecture for smart meter.

(UP) gives the bill details, while the sensors supply the data for meter readings. The bill details are subsequently sent to the WS and ultimately to the user.

7.2.2 Implementation of AMI

Data management systems, communication networks, and smart meters are all integrated into AMI. Across millions of utility customers served by local or national utility providers, this integrated system is in place. Three critical factors for the implementation of AMI involve connectivity, service equipment power consumption, and hardware implementation costs. Firstly, it is essential to provide consistent connectivity between each consumer and the utility (AMI service provider). To maximize the efficiency of power utilization, optimal communication is necessary. Secondly, while the power consumption of individual user equipment may seem insignificant, the aggregate impact across a large user base becomes significant. Reducing equipment power consumption contributes to overall efficiency. Thirdly, the economic sustainability of the AMI implementation depends on controlling hardware and installation expenses, particularly when implementing the system for a large number of users. Thus, high connectivity, low power, and low-cost advanced metering infrastructure (HiLL AMI) are essential to address all these issues.

One design principle to guarantee cheap costs and good dependability is to leverage existing established technology. This strategy lowers the risk involved in implementing new, unproven components by leveraging established technology. Typical signal connection issues between user smart meters and the utility's AMI communications network access point are intended to be resolved by the modular construction of components. This flexibility improves the system's adaptability. For AMI to be implemented effectively, all user smart meters must be seamlessly integrated into the utility's communications network. A strong infrastructure is built on this close connection.

The smooth operation of the infrastructure creates a foundation for the implementation of several plans that enable the utility to produce and consumers to utilize power with maximally optimized efficiency. A major objective of AMI is this optimization. Flexibility is provided by the linked infrastructure to utilities as well as customers. Users have the option, for example, to pay less for power used during off-peak hours. Utilities profit concurrently from a more stable grid load. The decrease of carbon emissions is one of AMI's main objectives. AMI allows customers to utilize power when there is less demand, which helps to reduce carbon emissions and make better use of available resources. Users' ability to help reduce carbon emissions under AMI is greatly facilitated by smart meters on their end. This is accomplished by keeping a steady network connection with the utility company. Users are able to actively engage in demand response programs and other energy efficiency efforts when they have a solid connection, which guarantees reliable data interchange.

7.2.2.1 Communication protocols

The power grid is seeing a significant increase in the penetration of distributed energy sources and renewable energy sources, which is driving up intra-state, inter-state, inter-regional, and international electrical energy transactions. Load dispatch centers are responsible for managing the effective scheduling, measurement of energy imports and exports, billing, and settlement of bulk energy transactions, regardless of whether the energy comes from traditional

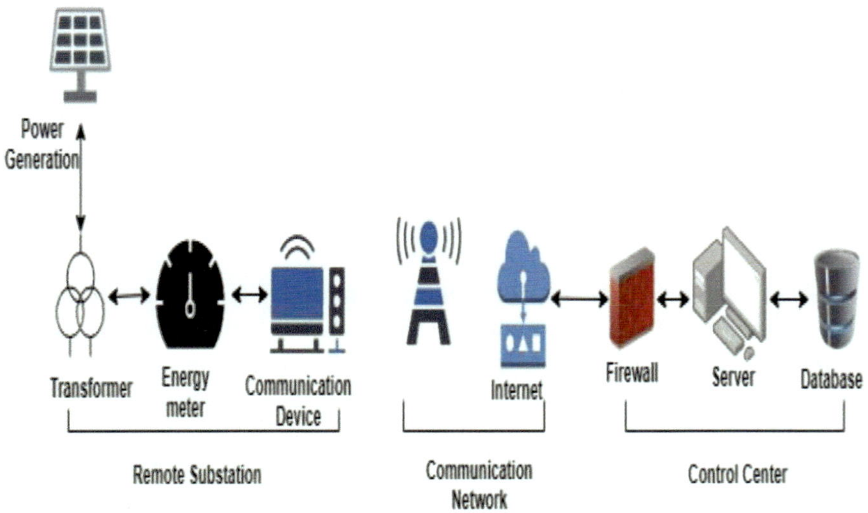

Figure 7.5 Architecture diagram of the AMI system.

or renewable sources. For effective scheduling, load/renewable forecasting, energy accounting, outage management, real-time deviation monitoring, billing, and transaction settlement, load dispatch centers rely on meter data.

The advanced metering infrastructure (AMI) includes the meter data acquisition system (MDAS) [4], which is responsible for automatically gathering meter data from substation interface energy meters and sending it to a central database for power system operations. Utility companies use this technology primarily to monitor in real-time and forecast future generation or consumption based on historical and current data. Utility companies and consumers can both benefit from real-time meter data information by using it to make more efficient use of electricity.

Meters, data concentrators, optical fiber networks, GSM/GPRS/3G/4G, power-line communication carriers, wired and wireless network components, and control center IT infrastructures are all included in AMI technology. When meter data is unavailable due to a fault or malfunction in the AMI system, power system operations are negatively impacted. The control center and meter communication are also open to different types of intrusions. Load dispatch centers must prioritize high availability, dependability, and integrity of meter data to manage the grid effectively.

Figure 7.5 displays the general architecture diagram of the AMI System. The enhanced security profile, association requirements, different parameter lists, on-demand data services, push data services, firmware upgrade services, and features must all be supported by smart energy meters [5, 6]. Transformer-operated HV/LV consumer meters must adhere to the set of parameters for category D-3. Boundary/bank/ring/availability tariff meters operated by transformers must adhere to the D-4 category parameter list. Numerous profiles, including immediate, block load, daily, billing, event, nameplate, and programmable parameters, are included in the parameter list [5].

Every sub-station has a data concentrator unit (DCU)/communication device installed, which serves as an interface between control centers and sub-station IEMs. In accordance with the configuration that the control center provides, DCU obtains data and status from IEMs at predetermined intervals and transmits it to the control center. Similarly, the control center provides DCU with control commands that it executes on IEMs, such as time synchronization, connect/disconnect, configuration changes, firmware upgrades, etc.

Through ethernet, DCU gathers data from IEMs and transmits it to the control center. The AMI system uses a variety of communication channels, including wired Internet, radiofrequency, optical fiber communication (OFC), power-line carrier communication (PLCC), GSM/GPRS, and others. The information gathered at the head-end system facilitates the tracking of energy flows in the grid network, supporting energy audits, monitoring, and decision-making as well as consumer energy usage, invoicing, notifications, and the detection of tampering and outages.

Real-time meter data loss can result from a number of AMI system defects, including failures in the head-end system, DCU, and communication network. There are a number of potential attacks on the communication between the control center and the meters. Operations of the power system are affected in a cascaded manner by communication attacks and data loss. The suggested solution aids in detecting the location of failures, boosting high data availability, and facilitating safe communication between control centers and IEMs. The main specific protocols used in these systems are as follows:

- ANSI C12: This protocol establishes guidelines for North American data collection systems and electricity meters' communication interface. It guarantees compatibility between various data management systems and metering devices.
- DLMS/COSEM: For smart meter communication, DLMS/COSEM is a widely used worldwide standard. Utilities can gather consumption data

and carry out remote management tasks thanks to the architecture it offers for data exchange between meters and data collecting devices. An association is the initial stage of meter data communication, according to protocol. During the association connection formation step, authentication takes place. There are two formats used for data transmission between the meter and the client: plaintext and ciphertext. Simple text messages are simple to read and can be easily intercepted by someone looking to listen in on you, which might cause privacy issues. When an attacker successfully executes a man-in-the-middle attack and gains access to protocol details, they can alter or tamper with the meter data, which can cause issues with data integrity.

- IEC 62056: Also referred to as the "IEC 61107" or "IEC 62056" protocol, this standard outlines the protocols for exchanging meter data across a range of communication channels, including serial interfaces and optical ports. It makes it possible for data-gathering systems and meters to communicate.
- Zigbee: This is a low-data-rate, low-power wireless communication protocol that is frequently used to link smart meters to smart appliances, in-home displays, and other consumer electronics. It works in buildings or residences and is appropriate for short-range communications on the 2.4 GHz frequency spectrum.

7.2.2.2 Radio frequency (RF) mesh

Smart meters that are networked together to form a mesh topology make up RF mesh networks. As a node in the network, every meter transmits data to other surrounding meters in order to arrive at a central collection point. Due to its scalability and resilience, RF mesh communication is appropriate for extensive installations in suburban and metropolitan settings.

There are two types of radio frequencies: RF cellular and RF noncellular. RF Mesh and RF star/point to multi-point comprise the noncellular radio frequency in this study. Because of RF mesh, any receiver can also function as a transmitter. Each radio frequency point serves as a frequency-hopping device to advance and boost transmitted signals. Wi-sun, Zigbee, and RF Silver Spring Network (SSN) are examples of RF mesh, whereas Lora, Wi-Max, Zwave, and Sigfox are examples of RF stars.

Since data is superimposed on frequency in wireless communication, frequency is crucial. Higher frequency in radio frequency corresponds to higher bandwidth. It will narrow the communication's scope. For this reason, line of sight (LoS) is more important for RF stars than for RF mesh. Two

Table 7.1 ISM band frequency.

Frequency range		Center frequency	Bandwidth
13.553 MHz	13.567 MHz	13.56 MHz	14 kHz
26.957 MHz	27.238 MHz	27.12 MHz	326 kHz
40.66 MHz	40.7 MHz	40.68 MHz	40 kHz
902 MHz	928 MHz	915 MHz	26 MHz
2.4 GHz	2.5 GHz	2.45 GHz	100 MHz
5.725 GHz	5.875 GHz	5.8 GHz	150 MHz
24 GHz	24.25 GHz	24.125 GHz	250 MHz

types of frequencies are utilized in conversation. The first is the ISM band, a region of the radio spectrum where RF energy is used for telecommunications as well as business, science, and medicine. Without a license, this frequency range is useful for a variety of uses.

Devices operating on ISM bands must be ready to tolerate interference from other ISM devices, according to the International Telecommunication Union Radiocommunication Sector's (ITU-R) spectrum regulation. Table 7.1 displays the ISM band frequency [7].

7.2.2.3 Power line communication (PLC)

PLC uses the power lines that are already in place to communicate with data-gathering systems and smart meters. Although it does away with the need for extra communication infrastructure, it might not work well in places with a lot of interference or bad power lines.

In general, PLC is the medium used by, a communication technology. Two varieties of PLCs have been utilized recently, based on frequency: broadband PLC (BB PLC) and narrowband PLC (NB PLC). While BB PLC operates in the frequency range of greater than 1 MHz, NB PLC operates in the 3–500 kHz range. NB PLC has matured to the point where it is frequently used with smart meters [7]. The two band standards that are most frequently utilized with NB PLC are the FCC (US standard) and CENELEC A (European standard). Low data rate connections (up to 250 kbps) are referred to as NB PLC, and they are very helpful for switching and monitoring [5]. The orthogonal frequency division multiplexing (OFDM) technique is used to address several issues, including latency, narrow bandwidth, error correction, etc. The standard protocols Prime and G3 are frequently utilized in NB PLCs. With many error-correcting techniques, G3 is more resilient than Prime. Prime, however, offers a faster data rate.

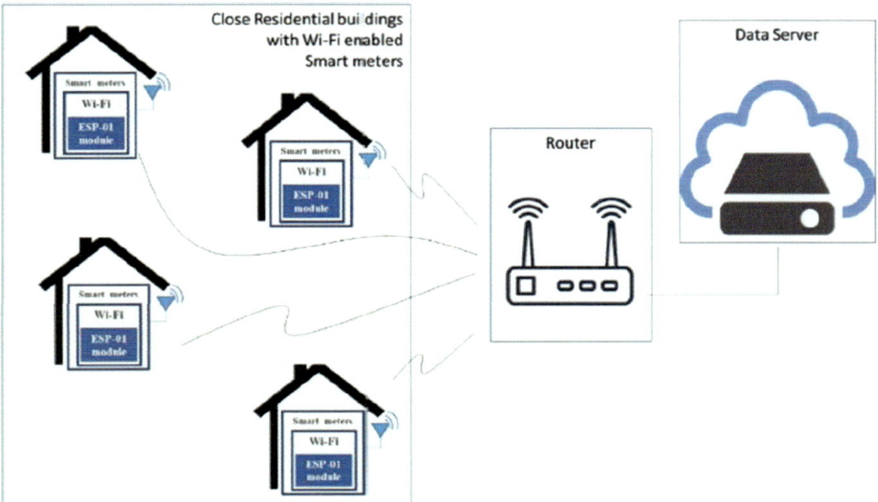

Figure 7.6 Wi-Fi network architecture

7.2.2.4 Wi-Fi

For data transfer, smart meters equipped with Wi-Fi modules can establish a direct connection to a nearby Wi-Fi network. This method is frequently applied to in-home energy management systems, which allow users to track their energy use in real time through online portals or smartphone apps.

An ESP-01 module was utilized in the implementation of the Wi-Fi communication network. The most well-known producer of Wi-Fi modules, Espressif Systems, created the ESP 8266 series, which was taken into consideration while choosing this particular Wi-Fi module. Based on operational characteristics such as form factor, power demand, the functional requirement of the smart energy monitoring circuit, and cost, several modules based on the ESP 8266 SoC were taken into consideration and compared. The modules that are taken into consideration include NodeMCU, ESP-01, ESP-05, ESP-12S, and ESP-201. The most widely used and affordable Wi-Fi system on chip (SoC), known as ESP-01, is made by Espressif Systems and features a low-power 32-bit microprocessor and the TCP/IP (transmission control protocol/internet protocol) stack [6]. If necessary, it can be integrated via the universal asynchronous receiver-transmitter (UART) interface with the application processor – in this case, the Atmega328p. Figure 7.6 shows the implementation of the Wi-Fi network [8].

7.2.2.5 LoRa and Low-Power Wide-Area (LPWA)

The Low-Power Wide-Area (LPWA) family's exclusive technology is called Long Range (LoRa). Its hierarchical structure and extremely basic wireless infrastructure make it suitable for a wide range of distributed measurement and Internet of things applications. Because of its star topology, the network makes it easier for end nodes to connect with base stations and gateways that gather and forward messages from the Internet to one or more network servers. User data that is transferred to or received from the end device is secured by the application server. It is important to note that LoRA/LoRaWAN only addresses the physical layer and the data link layer. It does not cover the technology utilized in the network backbone.

Several businesses and IT industry giants are presently utilizing a powerful marketing approach in the smart cities space to make their proprietary technologies available for usage or future development. Getting a marketing position and drawing in new customers is the major objective here. No manufacturer offers a complete LPWA solution as of yet, not even one that uses LoRa technology. The various players' suggested fixes are neither mutually exclusive nor a component of a single "defacto standard." Because of this, prospective users are still unsure about the advantages of each LWPA solution and which one best suits their requirements. This is the situation with LoRa for smart metering, where there are a lot of characteristics, and establishing a trial site will be expensive. Beyond what is currently theoretically available to the public, a performance review may be able to verify whether the LoRa solution can technically meet the needs of smart metering. We will be able to eliminate confusion and any expensive or time-consuming procedures before the deployment in this way [8]. Major areas can be described separately as below;

- **LoRaWAN**:

LoRaWAN is a low-power, wide-area networking technology intended for low-data-rate long-distance communication. It works well for long-distance smart meter connections in isolated or rural locations with spotty cellular service.

- **NB-IoT/LTE-M**:

Cellular communication technologies like LTE-M and narrowband Internet of things (NB-IoT) are designed with low-power, wide-area Internet of things applications in mind. Compared to conventional cellular technologies, they offer better battery life and more extensive coverage for smart meter connectivity in urban and suburban settings.

7.3 Smart Energy Management Systems

Energy management systems (EMSs) are considered essential for grid operation. The EMS performs several essential functions for the microgrid which include:

1. Economic dispatch: Economic dispatches are algorithms that optimize the use of resources in the utility network economically while maintaining the system constraints. In the economic dispatch algorithm, the scheduling of DERs such as ESS, charge/discharge schedule for EVs, generation/load curtailment as well as demand response decisions.
2. Unit commitment: Unit commitment algorithms determine which generating units such as gas turbines and diesel generators are to be turned on in the power system or the microgrid. This algorithm optimizes the scheduling of units in terms of cost while maintaining the generator limits, ramp rates as well as maintenance schedules of units.
3. Optimal power flow: Optimal power flow algorithms optimize the power flow in the transmission system according to the security constraints which include line capacity limits, voltage limits, line losses, and so on.
4. Voltage regulation: The system voltage is maintained by controlling devices such as synchronous machines, capacitor and inductor banks, transformers, and flexible AC transmission systems (FACTS).
5. Frequency regulation: The system frequency is maintained by controlling the generating sources and the loads.
6. Load forecasting: Since economic dispatch algorithms are reliant on knowing the load which may always not be known ahead of time, load forecasting models help address the uncertainty associated with the required load during the day.
7. Renewable energy forecasting: Since renewable energy is also inherently variable, the economic dispatch algorithms can make use of forecasting models to determine the system state and make decisions accordingly.
8. Energy market participation: The EMS can act as a gateway for prosumers to participate in the electricity market which includes the wholesale electricity markets, ancillary service markets as well as the capacity markets (spinning reserves).
9. System stability and security: The EMS ensures that the microgrid and grid operate under stable conditions and can identify and clear faults in the network quickly. At the same time with the deployment of IoT devices, the EMS also needs to ensure data security, data privacy, and safe transactions between participants.

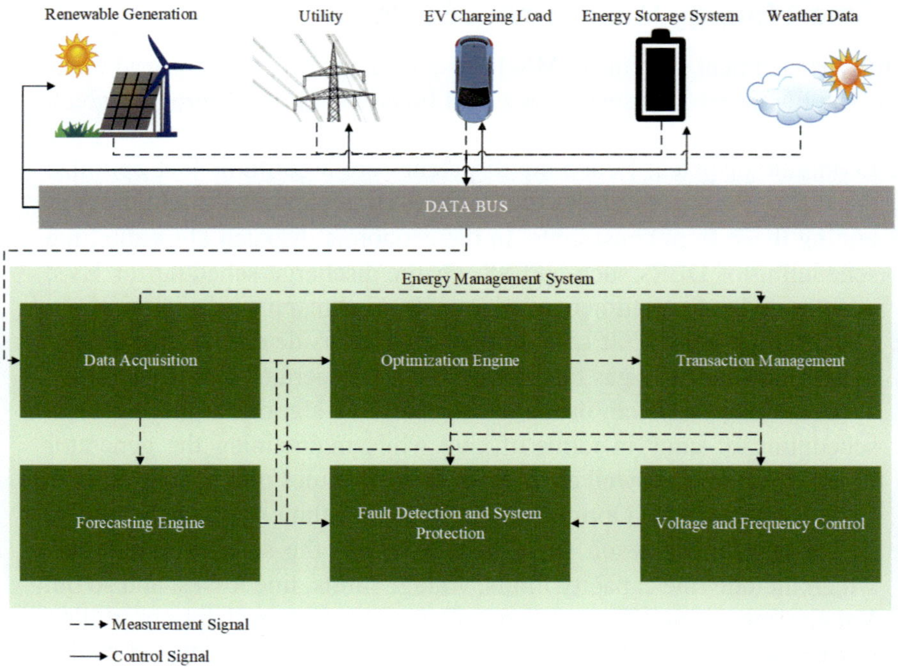

Figure 7.7 Energy management system framework.

As can be seen, some of the functions are interlinked and the included functions can vary based on where they are placed in the power system. They can be placed in the transmission system, distribution systems, microgrids, or nanogrids. A lot of the above-mentioned functions can be broadly categorized into different categories as depicted in Figure 7.7.

In the EMS, firstly all data from all the players in the system are gathered through the implementation of smart meters and sensors. These measurements include EV data, power demand, weather information, renewable generation, total generation mix, electricity prices, and so on. Once the data is received, the data acquisition system sends the data according to the functions that are needed to be performed. The weather information is used by the forecasting engine to forecast the power demand as well as the renewable generation. The forecasted data along with the real-time data is then sent to the optimization engine to perform functions such as demand response, load curtailment, generation curtailment, economic dispatch, unit commitment, or optimal power flow depending on the location in the power system. Based on the optimized power system operation, the transaction management system

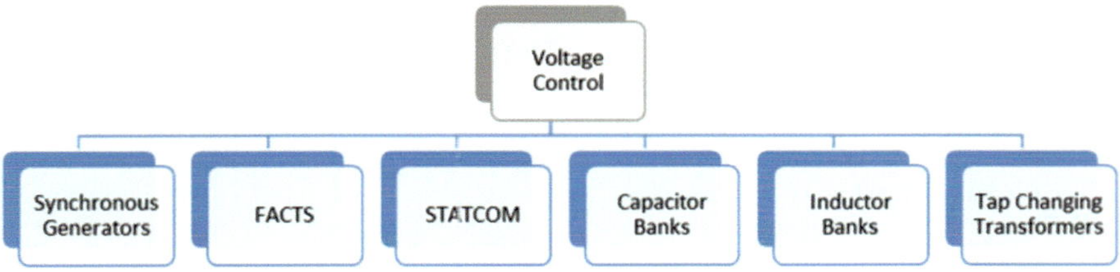

Figure 7.8 Different devices to control the system voltage.

takes care of handling the accounts and the billing among the different stakeholders.

The voltage and frequency control system controls the system voltage and frequency through different techniques. The system voltage is mainly controlled by controlling the reactive power flow in the system while the system frequency is maintained by controlling the real power flow in the system. The devices used to control the system voltage are shown in Figure 7.8.

The system frequency is controlled through the system inertia present in the system which is determined by the power flow in the system. There are three levels of frequency control in the system: primary, secondary, and tertiary. The system frequency changes when there is a mismatch between power generation and consumption. The primary control system performs actions to remove this unbalance through actions such as increasing the generation. The secondary control system then kicks in to restore the system frequency to its original value as the primary control system is incapable of doing so. This is achieved through the usage of reserved power. The tertiary control system then tries to restore the reserves used in the secondary control. The frequency plot versus time and the impact of the control actions in each time step is shown in Figure 7.9.

However, given the interconnected nature of power system due to geographical considerations, the impact of increasing or decreasing the power generation based on frequency changes needs to be taken into account. To do so, the area control error (ACE) parameter is used to determine the generation from the different resources in the system. Also, the ramp rates of the generating units play a major role in controlling the system frequency. Although, the control techniques developed above work well when there are synchronous machines in the system to generate the inertia required for the power system to operate, with the increasing penetration of renewables and inverter-based resources, obtaining a similar performance has become

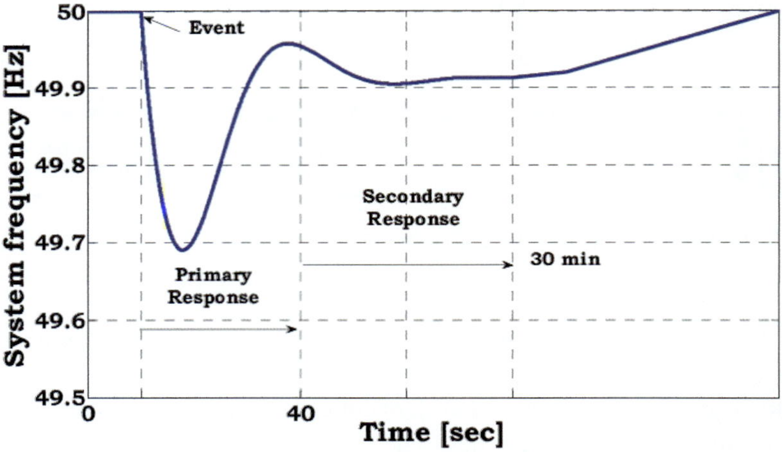

Figure 7.9 System frequency change and response to control action.

challenging. In systems with inverter-based resources (IBRs), the concept of virtual inertia through grid-forming inverters is currently being investigated.

Based on all the data acquired and system information received, the fault detection and protection system block continues to monitor all the system parameters such as voltages and currents across the network to identify the type of fault, and fault location and isolate the system from the fault location. Since a lot of these functions require many computations, technologies such as cloud computing and machine learning models play a major role. In addition, to ensure that the transactions between the stakeholders occur in a safe, reliable, efficient, and secure manner, the use of blockchain technology is beneficial for EMS.

For example, in [9], using blockchain, the EMS determines charging/discharging schedules for EVs and the subway transportation system. The blockchain is used for secure data transfer while the uncertainties associated with wind generation and the transportation system is modeled through a statistical architecture based on an unscented transformation algorithm. The authors in [10], have looked at using cloud computing for developing the energy management scheme based on electricity prices being dynamic.

Machine learning and deep learning models can also be applied to energy management systems in several ways. They can be used to forecast power demand and renewable generation as was done in [11]. They can also be used to develop economic dispatch algorithms as was done in [12]. The authors in [12] used deep learning to develop the energy management scheme while using the data from IoT-based sensors. A detailed review of the current work on the application of edge computing and cloud computing for smart grids has been presented in [13]. Other methods based on rule-based approaches, reinforcement learning, and game theory models have been presented in [14]–[16]. Along with the EMS, power electronic converters play a major role in today's power network. They help interface the renewable generation with the power grid and regulate the system voltage and frequency to match the grid voltage and frequency.

7.4 Power Electronic Converters for Grid Integration

Power electronic converters, which enable the highly efficient and controllable conversion of electrical power from one form to another, are crucial parts of electrical systems. These converters adjust the electrical signals' voltage, current, and frequency, providing a variety of features in contemporary electrical systems. The different types of power electronic converters are summarized in Figure 7.10.

Power electronic converters are mainly categorized into four different categories: 1. AC to DC Converters, 2. DC to DC Converters, 3. DC to AC Converters and 4. AC to AC converters. AC–DC converter devices change alternating current (AC) into direct current (DC). Also, they are named as rectifiers as their function is to rectify the AC waveforms. These devices change alternating current (AC) into direct current (DC). They are frequently

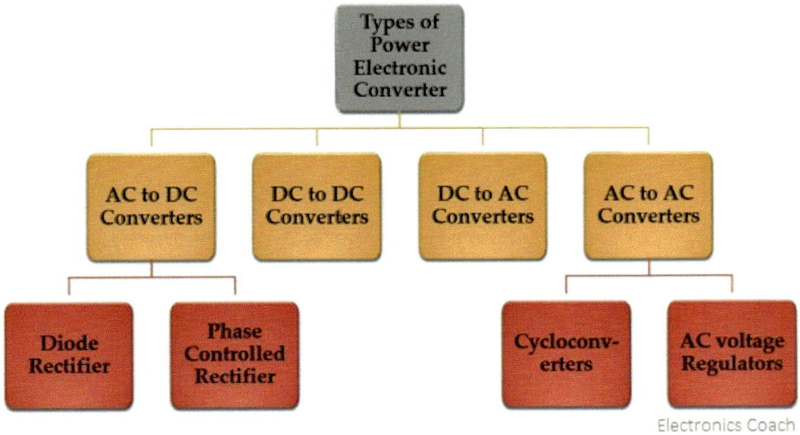

Figure 7.10 Types of power electronic converters.

utilized in applications including variable frequency drives, battery chargers, and electronic device power supplies. Figure 7.11 shows the simple AC–DC converter diagrams.

DC–AC converters can be identified in several applications. Direct current (DC) is changed into alternating current (AC) using inverters. They are extensively utilized in grid-tied systems for redistributing power back into the grid, electric vehicle powertrains, uninterruptible power supplies (UPS), and renewable energy systems (such as solar and wind). Figure 7.11 shows the structure of the simple inverters which can be mainly categorized as single-phase and three-phase inverters depending on the application.

DC–DC converters can be identified as the converters that can convert the power levels of the same type of voltages. These converters convert between different DC voltage levels, either higher or lower. Applications for them include energy storage systems, battery charging, and voltage regulation. There are various types of DC–DC converters available depending on the

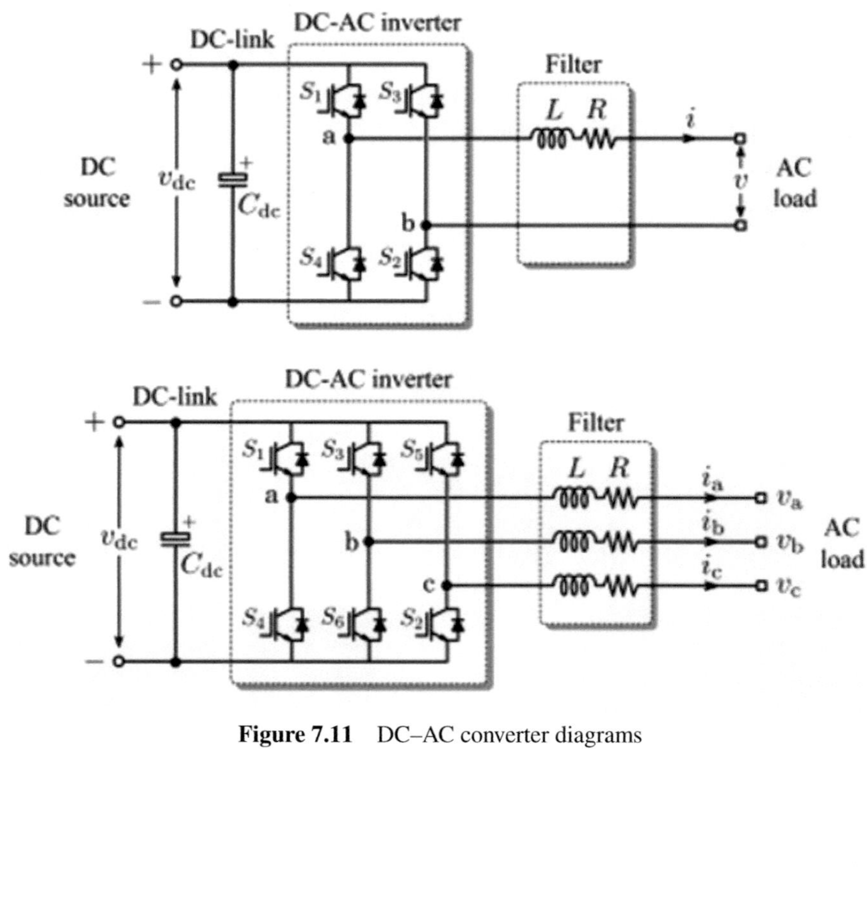

Figure 7.11 DC–AC converter diagrams

application. Buck converters, boost converters, and buck-boost converters can be identified as the major types of converters which are shown in Figure 7.12.

The AC–AC converters are also able to convert the same type of voltages. These converters also referred to as cycloconverters, directly change one AC voltage or frequency into another. They are employed in applications like HVAC systems and AC motor speed control that call for variable frequency and voltage management. Figure 7.13 shows a type of three-phase AC/AC converter which can control both voltage and frequency.

The specific types of power converters can be found in the literature. Voltage source converters (VSCs) are one of them. Power electronic converters known as VSCs are utilized in flexible AC transmission systems (FACTS) and high-voltage direct current (HVDC) transmission systems. They provide effective power transmission and grid stabilization by offering precise

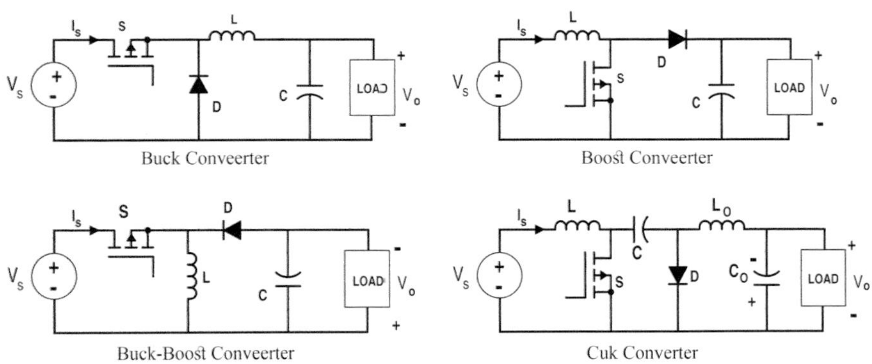

Figure 7.12 Main types of DC–DC converters.

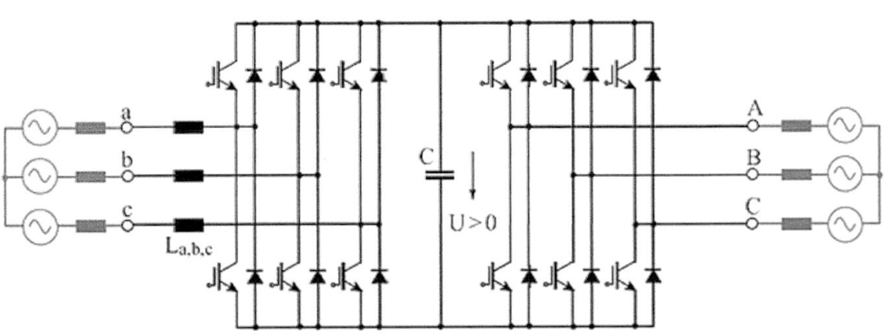

Figure 7.13 Three-phase AC–AC converter.

control over voltage and reactive power. Also, the current source converters (CSCs) are very common in power electronics. CSCs are utilized in HVDC systems similarly to VSCs. They are appropriate for applications needing high reliability and fault tolerance and offer a constant current output. The resonant converters are resonant circuit-based converters that are well-known for their low electromagnetic interference (EMI) and excellent efficiency. Applications for them include induction heating, wireless charging, and power supplies. The matrix converter is a modern type of converter designed recently. Bidirectional AC–AC converters known as matrix converters have the ability to convert AC power straight between two frequencies, negating the requirement for a transitional DC stage. Renewable energy systems and variable speed drives are two areas in which they are used. Soft starters and variable frequency drives (VFDs) are identified as very specific types that are used only in a few applications. These specific kinds of power electronic converters are employed in motor control applications. By gently increasing the voltage applied to a motor, soft starters lessen mechanical stress during the beginning. By altering the output voltage and frequency, VFDs regulate the speed and torque of AC motors. Overall, power production, transmission, distribution, industrial automation, transportation, and renewable energy are just a few of the industries that depend on power electronic converters to increase the dependability, efficiency, and controllability of their electrical systems.

Because they make it easier to convert variable renewable energy outputs, like wind and solar power, into electricity that can be easily incorporated into the grid, power converters are essential to the grid integration of renewable energy sources as can be seen from Figure 7.14. Here are a few popular varieties of power converters used for renewable energy grid integration.

For solar panels or wind turbines to produce direct current (DC) output that can be fed into the grid, they need inverters to produce alternating current (AC). Different kinds of inverters exist in power electronic applications. Central inverters can be identified as one of them. Utility-scale solar projects usually employ them. Before supplying it to the grid, they transform the DC power from several solar panels into AC power at one central site. The string inverters are frequently utilized in small-scale business and residential solar installations. An inverter, which transforms DC power into AC power, is linked to each solar panel or a small cluster of panels. Microinverters, as opposed to string inverters, are mounted directly on each solar panel and convert DC power to AC power at the panel level. They have benefits including better monitoring capability and higher energy generation.

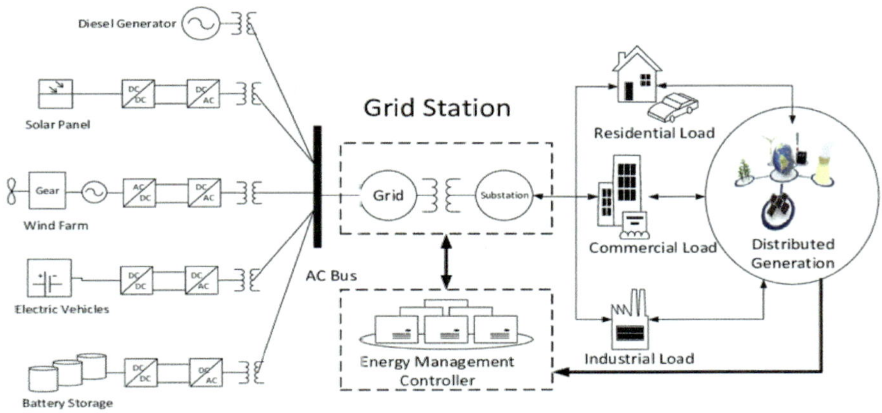

Figure 7.14 Sample architecture of a grid-integrated renewable system.

Voltage source converters (VSCs) are very common types used in systems. High-voltage direct current (HVDC) transmission lines, which are increasingly being used to transmit renewable energy over great distances, are where VSCs are employed. Through the conversion of AC power to DC for gearbox and vice versa, VSCs provide effective, low-loss power transfer over extended distances. When a renewable energy source's frequency differs from the grid's frequency, frequency converters are employed. They provide smooth integration by converting the renewable energy source's output frequency to coincide with the grid frequency. Grid-tie inverters with energy storage capability are an essential type of application in renewable integration. These inverters have features for storing electricity, like batteries. They store excess energy for later use or to provide grid support services like frequency management or peak shaving, in addition to converting DC power from renewable sources to AC for system integration. The controlling of a grid-tied inverter has major functions such as power control and energy management. Figure 7.15 shows a general block diagram of the control system of a grid-tied inverter with a closed-loop control system.

With the help of these power converters, renewable energy sources can be smoothly incorporated into the grid, increasing their penetration and lowering reliance on fossil fuels. Furthermore, to guarantee grid stability and dependability, sophisticated control and communication systems are frequently incorporated with these converters.

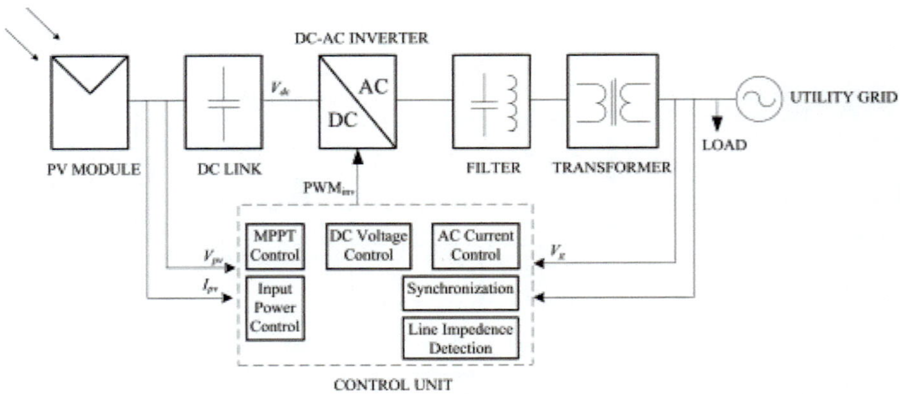

Figure 7.15 Controlling of a grid-tied inverter for PV integration.

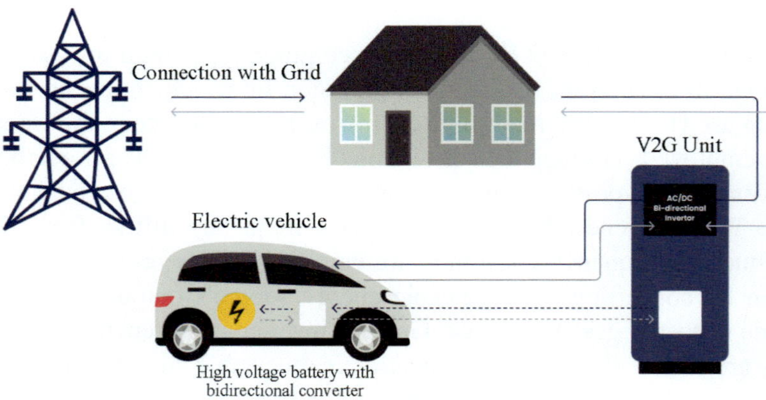

Figure 7.16 V2G application.

7.4.1 EV-grid integration for V2G application

The bidirectional flow of electrical energy between electric vehicles and the electrical grid is referred to as "EV to Grid" (V2G) integration as shown in Figure 7.16. Because of this idea, EVs can supply energy back into the grid when necessary, in addition to consuming it. There are several functions of V2G integration and its possible applications.

The number of electric vehicles has increased rapidly in recent times, and as technology advances, this will likely have a positive effect on the development of EVs across the nation soon. Climate change and worries about environmental pollution have been the main drivers of the sharp increase in the popularity of low-carbon emission automobiles. Pure electric vehicles (EVs) have become the most environmentally beneficial modes of transportation in recent times due to their lower carbon emissions. Soon, V2G technology will be essential to preserving a suitable, stable grid. The reactive power compensation of a smart grid using a bidirectional converter in the V2G application is covered in this study. Despite the benefits to the environment, concerns about grid overcrowding and potential instability are being raised by the potential growth of the electric vehicle (EV) industry. Further research on vehicle-to-grid technology is necessary to support the rise of electric vehicles (EVs) and increase the utilization of EVs while reducing the impact on the grid.

Infrastructure for charging that can transfer power in both directions is needed for V2G integration. Usually, they are smart charging stations with bidirectional chargers – that is, chargers that can converse with the grid as well as EVs.

As a specific application of V2G, the grid-supporting capability can be identified which they can act as distributed energy resources. When electricity rates and demand are low, EVs can be charged during off-peak hours. During peak demand periods, they can discharge power back into the system, lessening the burden on the grid. Through the adjustment of charging and discharging patterns in response to real-time grid circumstances, V2G systems can aid in the distribution of load throughout the grid. By taking up extra power from the grid during high-frequency events and returning it to the grid during low-frequency ones, EV batteries can help regulate frequency. EVs can contribute to the regulation of voltage levels in distribution networks, enhancing grid stability, by either delivering or absorbing reactive power.

The large-capacity battery of the EV can feed energy back to the SG to increase the stability and dependability of the grid in addition to converting electrical energy for mechanical propulsion. Vehicle-to-grid, or V2G, is the term for this reversal power transfer from EVs to SG, which is facilitated by both the quick commercialization of EVs and the advancement of batteries. According to International Energy Agency (IEA), there will be 56 million EVs on the road by 2030, with an average battery capacity of 120 kWh [17]. Since parking lots are where 90% of vehicle operation time is spent immobile [18], SGs can make optimal use of them while parked.

Grid resilience is one of the main advantages which is coming with V2G. Integrating V2G can improve grid resilience by offering backup power in case of emergencies or outages. When the grid is down, EVs can act as mobile energy storage devices, giving electricity to vital infrastructure or residential areas. By storing extra energy produced during times of strong renewable generation and returning it to the grid during times of low renewable output, V2G systems can aid in the integration of renewable energy sources. By taking part in grid services programs or selling extra energy back to the grid, EV owners can get money. This can assist in defraying the cost of charging and owning a car. Furthermore, there are lots of environmental benefits due to V2G applications. Through the increased use of renewable energy and the decreased requirement for fossil fuel-based plants, V2G integration can help lower carbon emissions and enhance air quality. Market mechanisms and supportive regulatory frameworks that guarantee equitable pay for grid services rendered by electric vehicles (EVs) and encourage participation are necessary for the successful implementation of V2G integration. Therefore, energy management will be a crucial factor in V2G for optimized energy usage as well as customer satisfaction. There are several topological methods in energy management based on V2G for charging stations and domestic charging stations.

As illustrated in Figure 7.17, there are three different V2G working modes depending on the scheduling target. Vehicle-to-home (V2H) mode is activated when EVs are connected to homes to supply energy for equipment designed

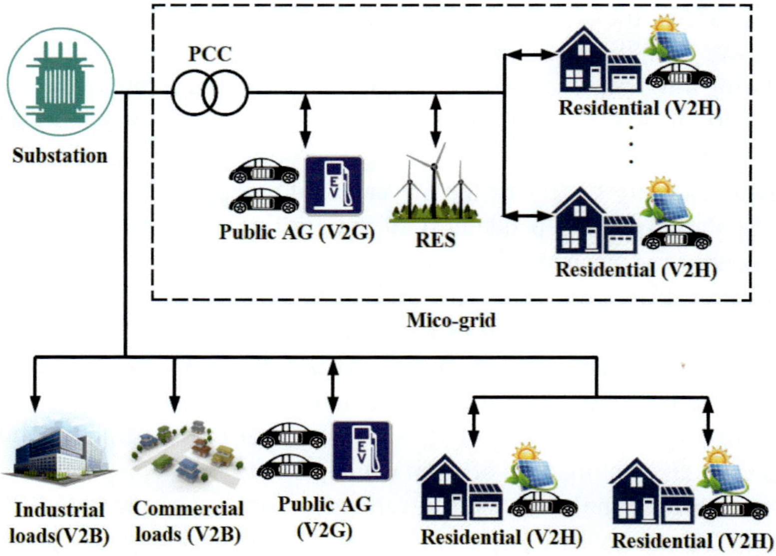

Figure 7.17 An overview of energy management of V2G.

for residential use with RES. EVs can operate in vehicle-to-building (V2B) mode to conduct discharging services when they are driven to places of employment (industrial loads) or retail centers (commercial loads) and linked to parking lot chargers. Vehicle-to-grid (V2G) is carried out at AG at peak hours to prevent overload.

All things considered, V2G integration has the power to revolutionize how we use and manage energy resources, improving grid sustainability, dependability, and flexibility while benefiting EV owners and society at large financially. However, removing market, legal, and technological obstacles will be necessary for broad adoption.

7.5 Conclusion

Advancements in various technologies like Internet of things, communication networks, blockchain, cloud and fog computing, machine learning, and power electronics have played a pivotal role in transitioning from conventional power grids to smart, integrated systems. These technologies find applications in areas such as smart metering, energy management, electricity markets, and integrating EVs into the grid. This chapter examined how these technologies are evolving to facilitate the transformation of traditional grids into smarter, more efficient systems.

The integration of vital elements including cloud and edge computing devices, wireless sensors, smart meters, and advanced metering infrastructure is critical to the effective deployment of smart grids. But there are a lot of obstacles with this integration, especially when it comes to cybersecurity and data privacy. Therefore, in order to properly address these issues, it is imperative that we look at these technologies' advances as discussed in this chapter.

Smart grids are complex systems that need smooth data sharing and connectivity between several parts. Wireless sensors offer more data on environmental factors and grid functioning, while smart meters are essential for gathering data on energy usage in real time. Devices for edge and cloud computing make it possible to analyze data, which speeds up the decision-making process for grid management and optimization. AMI coordinates the functions of many components and guarantees effective energy distribution, acting as the backbone of the smart grid system.

Despite their advantages, these technologies' integration creates risks that might jeopardize grid operations' security and privacy as well as that of customer data. Smart grid systems are exposed to serious risks from cybersecurity threats including hacking and data breaches, which call for strong security standards and procedures. For those involved in the implementation and maintenance of smart grids, a thorough examination of these developments is vital. Stakeholders may proactively address cybersecurity threats and guarantee the dependability and integrity of smart grid infrastructure by remaining up to date on the most recent developments and best practices.

In conclusion, the incorporation of smart grid technology has great potential to transform the energy infrastructure. Even if issues like cybersecurity and data privacy need to be addressed, there are workable solutions available thanks to continuous improvements in these fields. Stakeholders can confidently manage the shift to smart grids by using these advancements, guaranteeing a future in energy that is more safe, robust, and efficient.

References

[1] "Smart grids - IEA." Accessed: Feb. 19, 2024. [Online]. Available: https://www.iea.org/energy-system/electricity/smart-grids

[2] A. Dey, B. Chakraborty, S. Dalai, and K. Bhattacharya, "Insights and New Practices for Advanced Metering Infrastructure and Smart Energy Metering Framework in Smart Grid-A Case Study," *2022 IEEE Calcutta Conference, CALCON 2022 - Proceedings*, pp. 323–326, 2022, doi:10.1109/CALCON56258.2022.10060514.

[3] O. Kebotogetse, R. Samikannu, and A. Yahya, "A Concealed Based Approach for Secure Transmission in Advanced Metering Infrastructure," *IEEE Access*, vol. 10, pp. 84809–84817, 2022, doi:10.1109/ACCESS.2022.3195240.

[4] S. Dodla, L. Mahendra, K. Jaganmohan, R. K. S. Kumar, and B. S. Bindhumadhava, "Wireless Real-time Meter Data Acquisition System," *IEEE Region 10 Annual International Conference, Proceedings/TENCON*, vol. 2019-October, pp. 997–1002, Oct. 2019, doi:10.1109/TENCON.2019.8929650.

[5] M. Kisan and S. Sangathan, "Data Exchange for Electricity Meter-Reading Tariff and Load Control-Companion Specification [ETD 13: Equipment for Electrical Energy Measurement and Load Control]," 2011.

[6] "EXCERPT FROM Companion Specification for Energy Metering DLMS/COSEM Architecture and Protocols DLMS User Association," 2020.

[7] K. M. Tofani, K. G. H. Mangunkusumo, N. W. Priambodo, and B. S. Munir, "Consideration in Communication Media Selection for Advanced Metering Infrastructure in Indonesia," *Proceedings of the 2nd International Conference on High Voltage Engineering and Power Systems: Towards Sustainable and Reliable Power Delivery, ICHVEPS 2019*, Oct. 2019, doi:10.1109/ICHVEPS47643.2019.9011092.

[8] F. M. Dahunsi, H. Ijadunola, A. O. Melodi, and A. A. Ponnle, "Analysis of GSM, Wi-Fi and LPWAN communication technologies for Smart energy metering circuits," *Proceedings of the 2022 IEEE Nigeria 4th International Conference on Disruptive Technologies for Sustainable Development, NIGERCON 2022*, 2022, doi:10.1109/NIGERCON54645.2022.9803085.

[9] L. Zhang, L. Cheng, F. Alsokhiry, and M. A. Mohamed, "A Novel Stochastic Blockchain-Based Energy Management in Smart Cities Using V2S and V2G," *IEEE Transactions on Intelligent Transportation Systems*, vol. 24, no. 1, pp. 915–922, Jan. 2023, doi:10.1109/TITS.2022.3143146.

[10] A. Mondal, S. Misra, and A. Chakraborty, "Dynamic Price-Enabled Strategic Energy Management Scheme in Cloud-Enabled Smart Grid," *IEEE Transactions on Cloud Computing*, vol. 10, no. 1, pp. 111–122, 2022, doi:10.1109/TCC.2021.3118637.

[11] A. Golder, J. Jneid, J. Zhao, and F. Bouffard, "Machine Learning-Based Demand and PV Power Forecasts; Machine Learning-Based Demand and PV Power Forecasts," 2019.

[12] M. Abdel-Basset, H. Hawash, R. K. Chakrabortty, and M. Ryan, "Energy-Net: A Deep Learning Approach for Smart Energy Management in IoT-Based Smart Cities," *IEEE Internet Things J*, vol. 8, no. 15, pp. 12422–12435, Aug. 2021,doi:10.1109/JIOT.2021.3063677.

[13] J. Li, C. Gu, Y. Xiang, and F. Li, "Edge-cloud Computing Systems for Smart Grid: State-of-the-art, Architecture, and Applications," *Journal of Modern Power Systems and Clean Energy*, vol. 10, no. 4, pp. 805–817, Jul. 2022, doi:10.35833/MPCE.2021.000161.

[14] S. H. Hong and H. S. Lee, "Robust Energy Management System With Safe Reinforcement Learning Using Short-Horizon Forecasts," *IEEE Trans Smart Grid*, vol. 14, no. 3, pp. 2485–2488, May 2023, doi: 10.1109/TSG.2023.3240588.

[15] L. N. Tan, N. Gupta, and M. Derawi, "Adaptive Dynamic Programming and Zero-Sum Game-Based Distributed Control for Energy

Management Systems With Internet of Things," *IEEE Internet Things J*, vol. 10, no. 24, pp. 22371–22385, Dec. 2023, doi:10.1109/JIOT.2023.3303448.

[16] S. Chakraborty, G. Modi, and B. Singh, "A Cost Optimized-Reliable-Resilient-Realtime- Rule-Based Energy Management Scheme for a SPV-BES-Based Microgrid for Smart Building Applications," *IEEE Trans Smart Grid*, vol. 14, no. 4, pp. 2572–2581, Jul. 2023, doi:10.1109/TSG.2022.3232283.

[17] "Global EV Outlook 2017 – Analysis - IEA." Accessed: Mar. 10, 2024. [Online]. Available: https://www.iea.org/reports/global-ev-outlook-2017

[18] "Electric & Hybrid Cars: New Load, or New Resource? | Fortnightly." Accessed: Mar. 10, 2024. [Online]. Available: https://www.fortnightly.com/fortnightly/2006/12/electric-hybrid-cars-new-load-or-new-resource

8

Implementation of V2G and G2V Methods with Distributed Generation for E-Mobility

Kodeeswaran S[1], Kannabhiran A[2], and Nandhini Gayathri M[2]

[1]Department of Electrical Engineering, National Institute of Technology, Andhra Pradesh, India
[2]School of Electrical and Electronics Engineering, SASTRA Deemed to be University, India
E-mail: kodikrct@gmail.com; mail2akannan@gmail.com; nandhini.gayathri@gmail.com

Abstract

Electric vehicles charging infrastructure built with a piece of highly efficient equipment. Most of the vehicles were charged by plug-in charging methods. Due to safety, bulky cables, and uninterrupted power supply from the charger plug-in charging method has some limitations. To avoid such problems wireless charging methods play a vital role. Wireless charging for electric vehicle is an emerging research area that helps to reduce the charging time, cost and make as user friendly. At the same time, if the electric vehicles are more, this will produce power system problems in the generation and distribution sides. To avoid such power system problems smart charging methods (V2G & RES integration) will be highly efficient and to calculate the energy consumption, source from an electric vehicle to grid (V2G) supporting software's identified. This chapter explains the smart charging technology infrastructure developments and policies followed by different countries for clear understanding and promoting this technology further.

Keywords: Electric vehicles, wireless charging, renewable energy sources, policies, batteries, autonomous vehicles, e-mobility.

8.1 Introduction

Electric vehicle (EV) is a promising technology that does not emit greenhouse gases like CO, CO_2. These vehicles are driven by charged batteries. Due to the increase in number of EVs, it is very important to identify new charging technologies and power requirements. To deal with these issues, smart charging is identified as a very good and efficient technology. Power generating plants that are integrated with renewable energy sources are especially designed for efficient electric vehicle charging. In both the static and dynamic modes, the electric vehicle is connected by wireless chargers. During parking hours, these chargers are placed in residences, businesses, streets, and public spaces.

The power generation from renewable energy sources such as solar and wind that are interconnected to the grid for electric vehicle charging is nothing but smart charging. There are different modes used in smart charging, that is V1X, V2G, etc. It is a very efficient and low-cost method to use by customers. The V2G technology was first invented by Amory Lovins in 1995 and then William Kempton had developed this technique. During parking, the vehicle is charged, and also while not in use the fully charged battery is connected to the grid or home appliances. By this method, the customers can pay the amount at the lowest cost and sell the energy at a higher rate. To deal with this process it is very important to study smart charging technologies.

The smart charging technology may help the customers to pay the electricity bills. Also, the usage of electricity and the selling energy from customers i.e., V2G mode has enabled freely to pay the cost. There are some technologies like blockchain and other methods that connect the providers and customers. The smart charging technology enables renewable energy generation (solar and wind) in the distribution side. The energy generated by RES can be integrated into the grid. Based on this V2G technology, the customers interacted, and smart charging technology plays a vital role in the application of EVs.

This chapter helps the readers in understanding the basics and technologies used in smart charging. Section 8.2 explains about EV market and its growth and renewable energy integration for e-mobility. Section 8.3 deals with smart charging methods and different infrastructures. Section 8.4 presents vehicle integration to the grid and explains different business models. Sections 8.5 and 8.6 represent e-mobility outlook and global smart charging impacts.

8.2 State of Electric Vehicle Market

8.2.1 EV market evolution

With the scientific progress in the electrification of bicycles, cars, buses, and trucks and the growth in the market for these, electric vehicles are becoming more popular. In 2010, only about 17,000 electric cars were on the road. Figure 8.1 shows the country wise EV sales in the year 2019. By the end of 2019, 7.2 million cars had been sold worldwide, with China accounting for 47% and the United States for 20%. Around nine countries had over 100,000 electric vehicles on the road, with 20 countries having market shares of more than 1%. Global sales of electric vehicles are expected to reach 6.2 million by 2024, nearly three times the volume sold in 2019, according to S&P Global Market Intelligence.

Between January and December 2019, China sold 3.4 million highway-legal plug-in passenger cars, accounting for 47% of global sales. China also leads the world in the deployment of electric buses and plug-in light commercial vehicles, with over 500 thousand buses or 98% of the global stock, and 247.5 thousand electric light commercial vehicles or 65% of the global stock [1].

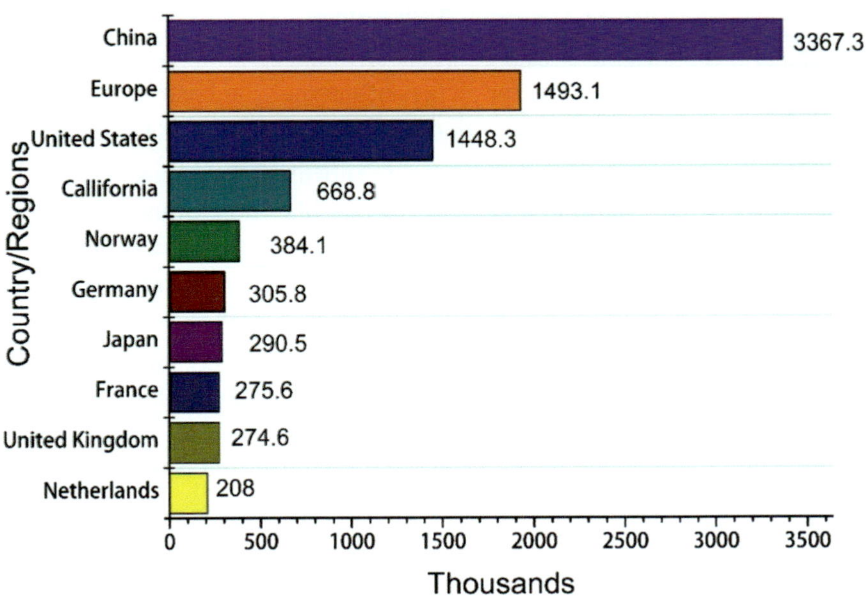

Figure 8.1 Country-/region-wise total electric vehicle sales up to 2019.

As of December 2019, there were 1.45 million plug-in cars on the road in the United States, accounting for 20% of global sales [2] [3], with California leading the regional market with nearly 670,000 plug-in cars sold [4–6]. Europe sold 1.7 million plug-in cars in 2019, accounting for 25% of global sales. Europe also has the world's second-largest stock of electric light commercial vehicles, accounting for 31% of the total [1].

With over 384 thousand light-duty plug-in cars sold by the end of 2019 [7], Norway is the most populous European country. Norway has the highest market penetration per rate in the world [8], with a market share of 55.9% of new car sales in the plug-in segment in 2019 [9]. In 2018, 10% of Norwegian roads were occupied by plug-in cars, rising to 13% in 2019 [10–11]. In 2019, the Netherlands has the highest density of electric vehicle charging stations in the world.

8.2.2 Support for charging infrastructure and policy incentives for electric vehicles

The government is encouraging the market uptake of electric vehicles by offering incentives and increasing the accessibility of charging facilities. These incentives, which fall into two categories: monetary and nonmonetary, have been used at the district, state, and local levels.

1) Financial incentives have resulted in tax advantages for owners of electric vehicles and higher taxes for owners of internal combustion engines. Germany, Norway, and the Netherlands all have similar policies.
2) Nonmonetary incentives serve as a viable substitute for expensive subsidies. To prevent traffic bottlenecks, nations like the US and Norway designate lanes for automobiles and buses. German towns and the UK have implemented an e-mobility toll road charging system in some zones to create low-emission zones.

Some governments like to implement these types of policies and set targets to increase the usage of EVs and reduce local air pollution [11]. Not only reducing the carbon monoxide emission, which deals with user benefits also. Various government policies and e-mobility targets were shown in Table 8.1.

In addition to policies, several governments set out to increase the number of charging stations because there aren't enough of them, which limits EV sales. During car parking, charging stations are installed in public spaces such as malls, offices, and other places. In order to put this strategy into action, the

Table 8.1 E-mobility targets and policy for different countries.

Country	Targets
China	The passenger EV car should increase up to 4% penetration in the market by 2020 ICE cars production and sale prohibition in the country possibilities were discussed "in the near future" 2017
France	As of 2040, ban on ICE cars sales
Germany	By 2020, planned to run 1 million EVs on the road From 2017 to 2020, targets to build 1000 new charging stations on the highway
India	As of 2030, ban on ICE cars sales
Japan	By 2030, EV sales should account for between 20% and 30% of total sales
Netherlands	By 2025, new gasoline and diesel cars will be prohibited
Norway	In 2025, only zero-emission cars and vans should be sold
Spain	Regulations are being developed that will prohibit the use of fossil petroleum vehicles by 2040 and their movement by 2050
UK	New petrol and diesel cars are prohibited from being sold, and 1.55 million electric vehicles are expected to be in use by 2040
US	By 2025, the city of New York hopes to have 20% of its residents owning an electric vehicle, and by 2040, it hopes to have 20% of its buses run on electricity. In 2025, the city of Los Angeles hopes to take a 10% part of electric vehicles, and by 2035, it hopes to have a 25% share. California states aim to reach 1.5 million vehicles in 2025 and 5 million by 2030, also the government planned to increase the charging station with the $2.5 million investment.

government must take the lead. As EV sales and business models increase, companies will be more inclined to provide and run charging infrastructure. The investments made by different nations to install charging stations are displayed in Table 8.2.

8.2.3 Renewable energy share in e-mobility markets

Renewable energy-based electric vehicle charging is an amazing and difficult technological advancement. Figure 8.2 illustrates how the grids, which are linked to the charging stations, were interfaced with wind and solar energy sources. An important part of the link between the grids and charging stations is power electronic components.

Figure 8.2 shows the block diagram of renewable energy sources interfaced with EVs charging stations. When the power is not generated by wind and solar systems, then the AC power grid will connect with charging stations [12]. Figure 8.3 shows the renewable power generation by various countries. In China and the US, these technologies play a major role in charging EVs.

Table 8.2 EVs charging infrastructure implementation.

Country	Program	Budget	Mechanism
Germany	10,000 level 2 and 5000 DC fast-charging stations will cost $285 million	$285 million	All eligible businesses will receive a 60% cost subsidy
Japan	Promotional project for the deployment of next-generation vehicle charging infrastructure Government-automaker collaboration in Nippon charge service	$1 billion	Local governments and highway operators receive grants Collaboration between the public and private sectors
Netherlands	Deal on the environment green	$31 million	On a case-by-case basis, contracts are offered to businesses
United Kingdom	Limits in residential charging station use Implementing DC fast-charging stations in major highways	$2 million $12 million	Municipalities apply for grants Grands and tenders administrated by a public body
China	New energy vehicle infrastructure incentive policy	-	Grants to provincial and local governments Local government subsidies for home and public EV charging
United States	The American Recovery and Reinvestment Act (ARRA) provides funding for public charging stations	$15 million	Grants for local governments with a matching component

In the US, "the Plugin EVeryWhere Network" has been developed by Austin Energy, which allows the customer to utilize 100% sources from wind generation for $4.17 per month. Approximately 35% of EV users utilized this benefit implemented by Austin Energy, also it offers to charge in-home during nights and weekends with additional low price. EVgo is another company that announces the charging plan by utilizing wind or solar energy sources. In 2012, the San Diego Zoo implements a charging station with 10 solar photovoltaic canopies with five charging stations. This idea can be utilized by customers, when not in use the energy stored in the batteries bank. Similarly, Google company also powers 750 charging points by renewable sources for their employees.

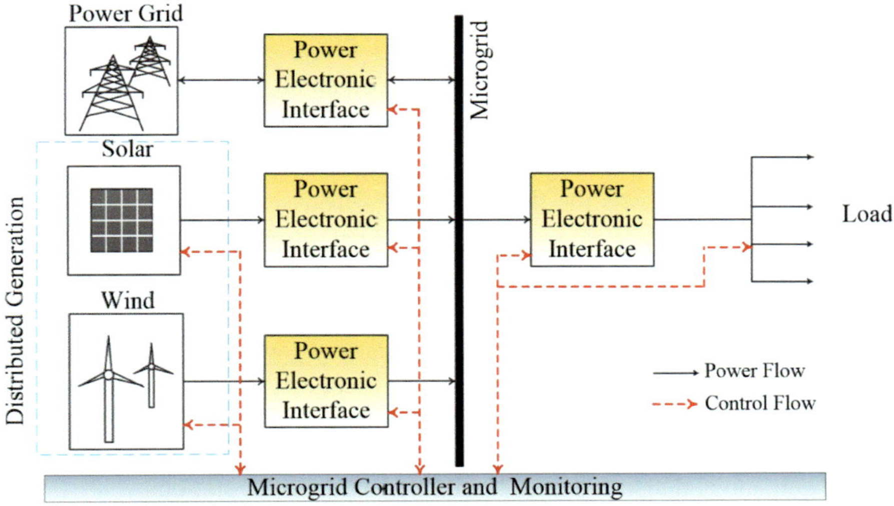

Figure 8.2 Block diagram of wind and solar interface.

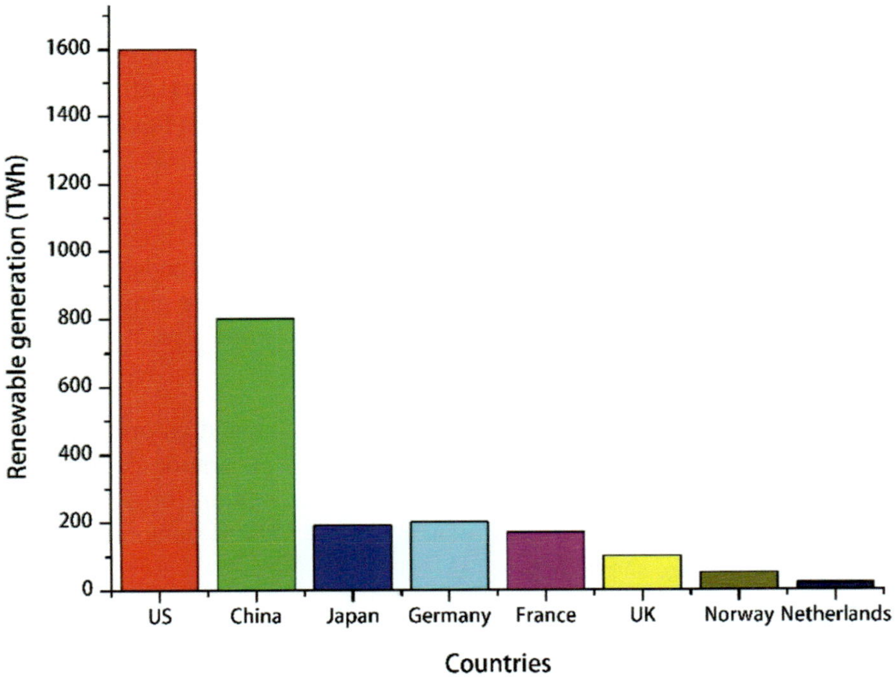

Figure 8.3 Renewable energy generation by countries.

8.2.4 EV flexibility potential

Today electric vehicles are very limited, and they have small batteries. The factors affecting electric vehicles' flexibility are given below.

1) The electric vehicle charging technology differs from slow, medium, and fast charging. It will take more time to charge the battery.
2) Battery capacity affects the flexibility of electric vehicles. 40–60 kWh battery with ranges 300 km running at one time fully charging is currently used.
3) Smart charging techniques are not used in normal charging stations.

To increase the EV flexibility potential and battery power connected from vehicle to grid by unidirectional (V1G) and bidirectional (V2G) techniques were identified to attract the customer with good benefits. During night time, parking a vehicle can supply power to the home and grid. This will reduce the power demand and also provides flexibility to the power stations.

In 2030, to increase EVs's flexibility high-penetration chargers will be available more in the market. Large-capacity batteries, approximately 90–200 kWh with a driving range of 600–1000 km, and smart charging stations will be built to reduce the charging problems as per International Renewable Energy Agency (IRENA) report. Also, it states that autonomous vehicles built from EVs will be more in the market in 2040. Between 2030 and 2050, the EVs impact will be high in the market and provide power system flexibility.

8.3 Smart Charging Outlook

This section provides an idea about different smart charging methods, infrastructure development status, ongoing research work, and existing projects.

8.3.1 Impact of charging EVs

Many EVs charging at a time results in increasing the demand on-grid side. Due to this reason, it is very important to upgrade the generation capacity [13]. The extension of generation includes a possible mixture of thermal, hydro, and renewable energy sources, penetration of EVs as demonstrated from various studies are shown below.

1) Impacts on electrical energy demand:
   In Europe, 100% of electric mobility scenario, EVs uses only 10%–15% of total generation. In Norway, 2.7 million cars only use 5%–6% of hydropower.

2) If the power demand by EVs is not distributed smartly, there will be an increase in the additional demand globally. For this problem, smart charging is a key. In the UK by 2035, the evening time charging of EVs will increase the demand by 3.5 GW. However, if smart charging is implemented which results in 0.5 GW demand only states by AER, 2018.
3) If not implemented smart charging results on the impact in local distribution grids. Four percent of distribution transformer will be overloaded by 5% of EVs. It is demonstrated by Xcel Energy, US.

Fast-charging methods require more power from the grid. For higher power, the charging cable may be bulky, charging station must support to maintain the same power. Also, it requires more expensive electronic devices on the vehicle side as well as on the charging side. This increases the cost and affects the feasibility. So, to avoid these problems smart charging is the best solution.

8.3.2 Smart charging

Smart charging is a method of optimizing charging based on the availability of grid and renewable energy sources, driver preferences, and electric vehicle supply equipment. Smart charging is not only used to minimize problems on the grid, but it should also be providing flexibility at the local and system level.

1) System flexibility includes wholesale market and transmission system power, which provides frequency control, voltage management, and power outages during an emergency.
2) Local flexibility includes distribution power and behind-the-meter optimization. Behind-the-meter optimization is used to increase the rate of renewable energy self-consumption of local production, lower the dependency of the grid, and reduce the electricity bill.

The availability of renewable energy sources makes it possible to charge EVs without using as much flexibility from conventional power plants. The degree of assistance for renewable energy integration is determined by the capacity and features of batteries used in electric vehicles. The majority of electric cars run on Li-ion batteries. Because smart charging technology can reduce costs while improving battery performance and being suitable for grid applications, it is a successful choice.

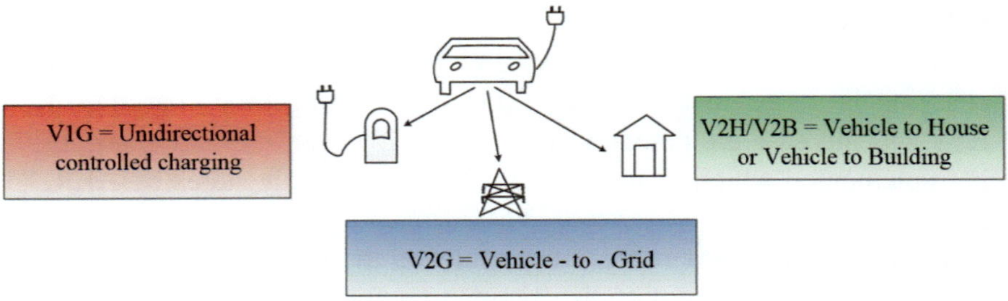

Figure 8.4 Smart charging formation.

Based on charging price and technical options it is classified as vehicle to everything (V2X) as shown in Figure 8.4. Specific configurations are:

1) unidirectional controlled charging (V1G);
2) vehicle to house or vehicle to building (V2H or V2B); and
3) vehicle to grid.

In each V2G operation, 1–2 MW should be exported from the vehicle. That means, 500 EVs with a power of approximately 3.7 kW connected to the circuit.

8.3.3 Charging infrastructure

The type of charging is very important to predict the feasibility of EVs. The EV batteries require power electronics circuits (AC–DC converter) to convert the sources from AC power grids. These converters are located in both off-board (charging point) and on-board charging (vehicle side). On-board charging method is used in level 1 and level 2 type chargers (up to 22 kW). To increase the quantity of charging solutions, distributors suggest DC charging or off-board charging method, which is available from 22 to 50 kW. The main charging locations are home, office, and public points. Mostly, 3.7 kW, 240 V AC chargers were used in the home. To increase the power by interconnecting solar productions, AC or DC charging methods were used based on requirements. DC power charging is normally implemented on highways. The mobility sectors prefer fast and ultra-fast charging methods; whereas, the slow-charging method is applied in smart charging techniques.

Tesla's ongoing research work states that in 2024, the driving range of 600 km should be achieved in one-time charging of electric vehicles. Electricity consumed by electric cars is about 20 kWh per 100 km, that means which requires 120-kWh battery for 600 km running. Compared to the 2010 range

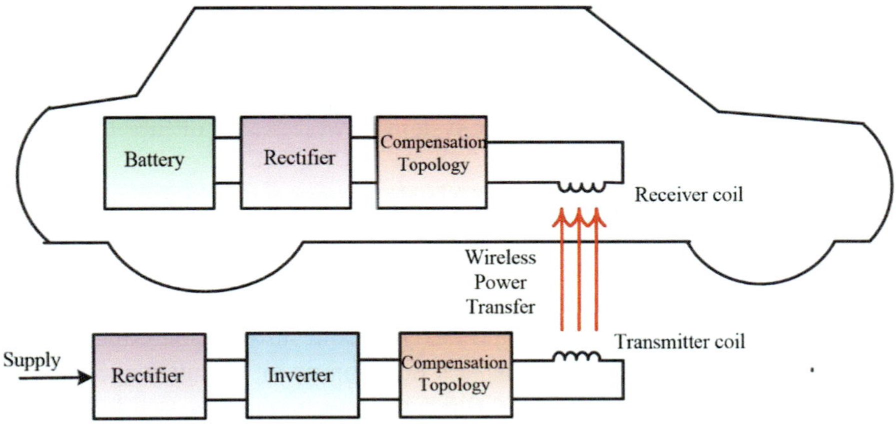

Figure 8.5 EV wireless charging infrastructure.

(20 kWh), it is six times higher. Also, this company focuses 200 kWh battery range for driving up to 1000 km. The power needed to charge a 200-kWh battery is 600 kW. The battery can currently charge at 3C, which means it can charge from 0% to 100% in 20 minutes. A 3C rate indicated that the entire battery will discharge within 20 minutes by discharge current. Since there are more EVs on the market, charging issues ought to arise more frequently. Thus, expanding and developing the charging sites is crucial. There are numerous efforts to expand the DC fast-charging infrastructure. This charging station requires a substantial upfront cost to construct. In the US, local towns and charging firms have formed an agreement to establish a workable solution.

In addition to wired charging, a quantity of different charging methods and revolutions are highly evolving. Wireless charging is a high-potential method that is already in use and will become more prevalent in the future. The static wireless charging method is one of the more advanced methods for improving wireless charging technology. The power output level from this charge for buses and trucks is 200 kW and for a car, it is 11 kW. Figure 8.5 shows the wireless charging infrastructure.

8.3.4 Smart charging enablers

Nowadays, the usage of electric vehicles is increasing significantly with improvement in the driving range. Once it reaches its peak in the market, the individual driver charging preference should be corrected to achieve flexibility. The customers who utilized smart charging have to be confirmed that the batteries are adequately charged for their travel. Even though there is a problem in terms of speed charging, the health of EV batteries and lifetime are accounted. To solve this problem, it is essential to design business models based on the requirements of the power system as well as that of the vehicle owners.

Big data and artificial intelligence (AI) algorithms can help to improve customer needs and acceptance [14]. Big data is used to interact with the grid to raise renewable energy shares. For example, home wall box charging is a keen charging technique that routinely charges the EV when energy costs are low. AI algorithms are used to find designs from information collected by sensors from the real world, to make a decision about maintenance and remote monitoring for different assets. Blockchain technology is used further to simplify the billing of EV services and provides flexibility. Blockchains are secured distributed ledgers enabling transactions. The advantages of black chains in EV charging are direct settlement, interoperability, and automation services.

8.4 Business Models and Regulatory Outlook

When electric vehicle adoption is more, the equipment supplier develops their new business model. In this section infrastructure models and e-mobility current position are going to be discussed in detail.

8.4.1 E-mobility performers

E-mobility includes:

1) EV sales include small EVs like bicycle, auto, and car are normal in price and are sold easily. For heavy vehicles and public transport subsidies or leasing are required to enhance the sale.
2) Mobility services include E-car sharing, fleet management, and charging stations.
3) Electricity sales for EVs charging and vehicle to grid connections.
4) Charging stations installation and maintenance, smart charging/billing systems.

The automotive industry seeks e-mobility to boost consumer confidence and ease anxiety; energy utilities shift their emphasis from supplying electricity to operating and maintaining charging infrastructure [15] to providing sustainable mobility; and oil and gas companies seek sustainable mobility. Consequently, these players seek mobility to maintain a profitable business

Table 8.3 Mobility actors positioning.

Current e-mobility value chain: examples of market actors				
	Electric vehicles (+ mobility services)	**Charging facility provision (installation/maintenance /operation)**	**E-roaming and replacements**	**Progressive services (V2G/Aggregation; EV + PV + storage; battery swapping)**
Original equipment manufacturers (OEM)	Tesla		⟶	• Smartphone • Superchargers
	Renault BMW Nissan Volkswagen Audi Mercedes Benz	*OEM's joint venture/partnership with governments, etc.* BMW Nissan Ford Mitsubishi Porsche AudiMercedes Benz	*E-roaming and e-roaming initiatives* BMW Bosch Siemens	• Storage combination • Energy management • Battery exchange • Smart home integration

and pass along the cost to their clients. The most important commercial models are delineated in Table 8.3.

8.4.2 Business models and V2G connections

Commercial models for EV grid connections were created using a wide range of ideas that are still in the early stages of development. These ideas include smart energy facilities, EV flexibility collection, battery exchange, and extended battery life.

8.4.2.1 Smart energy services supplier and collector

The commercial models for many loads and distributers were commercialized by monitoring and controlling many resources together. But EVs batteries aggregation process and services are not fully commercialized [16]. The key problem is profitability and competitiveness with other sources.

1) Price should be lowered by making solar generation with sufficient flexibility.
2) The EVs market volume should be increased to push the value lower.
3) Second-life EV batteries will be inexpensive.

8.4.2.2 Second-life storage applications

Second-life batteries are an alternate one for energy storage applications. Batteries should be replaced when their capacity reaches 70%–80%, i.e., it is not provided sufficient mileage but it's still in good form to use as a storage system. This will increase the lifetime of the battery to 10 years for $150 per kWh. During the low-cost period, it will be pre-charging from grid to battery and in the high-cost period it will be discharging, depending on the applications this may vary. Second-life batteries are mostly used in fast-charging stations, solar-powered streetlamps, telecom towers, smart home applications, etc. Battery swapping is another method, but it is greatly affected by many battery exchange difficulties of EVs fleets such as cabs need to be swapped several times in a day [17].

8.4.3 Guideline for V2G integration: Electrical energy markets

The expense of technology is impeding the adoption of EVs. Incentives for grid connections and market functioning should be offered to vehicles in order to lower the cost of technology. These kinds of cars need to be exempt from having to pay the cost. This owner should just cover the costs of net energy transferred. In retail marketplaces, the price of electricity should be set at a rate that is easily comprehended by customers [18]. Electricity incentives for rate plans and smart charging are already in place in some nations. EVs are regarded as a distinct load category for these designs. The price varies based on when it is used, such as during high and off-peak hours. In certain countries, such as the US, EV home charging is available at a 95% discounted rate at night.

8.5 E-Mobility Outlook

Fully battery-operated vehicles (BEVs), plug-in hybrid electric vehicles (PHEVs), and range extended electric vehicles (REEVs) are the different types of electric vehicles. Battery charging is applied in these types of vehicles from electricity. REEV is used in the non availability of electric charging, the small combustion engine used to charge the batteries. This section discusses the electric vehicle market and its evolution for a better understanding.

8.5.1 Cost and competitiveness of EVs

The main important issue that increases the cost of EVs is the battery cost and availability of batteries. In 2010, the battery cost was $1000/kWh and in 2017, it reduced to $200/kWh. Further, the cost is reduced, then electric vehicles are the best competitor to ICE vehicles. According to analysts, in 2025, the expected cost of the battery will reduce to $100/kWh. If it is happening, then the EV price will be low in the market and the number of customers will increase further. Also, a greater number of models' availability would reduce the cost of EV. Based on the driving range, battery capacity, and power train

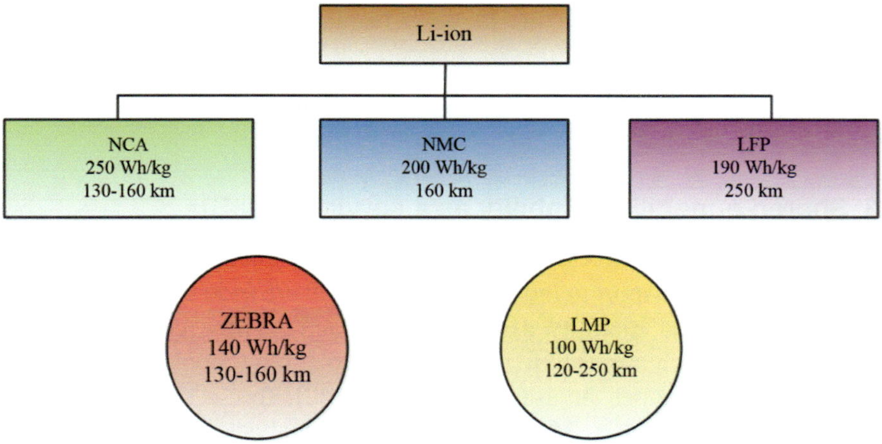

Figure 8.6 Real average data for Li-ion batteries.

the number of models reaching the market varies. This should reduce the cost of EVs and increase the numbers in the market [19]. The evolution of vehicle-related policies is an important concern. The policies may vary according to the region of countries. The government will take care of these policies to support the electric vehicle market.

8.5.2 Outlook for batteries

NCA – Lithium nickel cobalt aluminum oxide
NMC – Lithium nickel manganese cobalt
LFP – Lithium iron phosphate

The major factor that affects the cost of EVs is the cost of batteries. Most of the EVs are using Li-ion batteries to run the vehicle, at the same time Li-ion batteries have some limitations concerning safety and availability in the future, as well as cost. The real average data for Li-ion batteries is shown in Figure 8.6. The evolution of battery types not only affects mobility aspects which include driving range, fast charging, and the ability to connect to the grid. Two types of technologies were applied to develop the batteries to face the mobility challenges, i.e., Zeolite Battery Research Africa (ZEBRA) and lithium metal polymer (LMP). Also, some other technologies were under development and could be possible if the below issues were solved:

1) In Li-ion batteries, silicon should act as the –ve electrode.
2) Lithium Sulfur scheme.

3) Sodium-ion batteries have the higher advantage in terms of rate and ecological usage.
4) Metal air batteries (Al-air, Zn-air).
5) Redox flow batteries.

8.5.3 Shared e-mobility: Mobility as a service (MaaS)

Figure 8.7 indicates the process of MaaS step by step. Changing the owner-centered method of transport to mobility concepts that are used by all persons or customers is called mobility as a service (MaaS). These concepts are already existing, which are used for trip (or) tour planning, booking, riding, and payment of bills as per the usage of transport [20]. To meet the needs of customers on a timely basis, MaaS should require analysis, modeling, and data collection from the comprehensive operator portfolio. The simplified diagram for MaaS is shown below. MaaS providers allow the clear root to reach the destinations based on the customers to reach the locations, booking the trip, and bill generation in a single invoice. The customer's data and transport providers allow the MaaS suppliers to offer personalized knowledge.

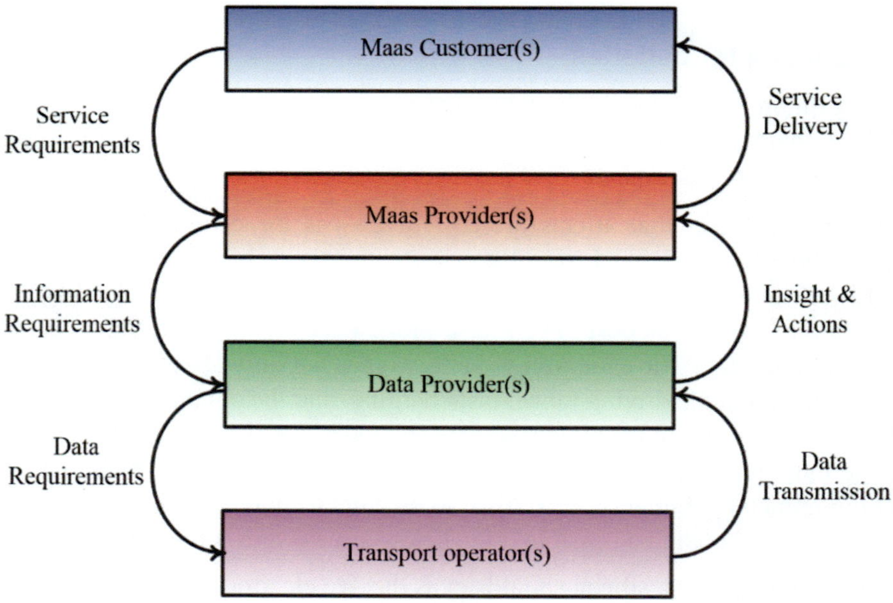

Figure 8.7 Chain diagram for MaaS.

8.5.4 Autonomous EVs

As discussed earlier, EVs emit fewer greenhouse gases, and the business managers mostly like EVs over ICE vehicles. Now, the autonomous vehicles are slowly entering the market as the business point of view for the growth of EVs and boost up the MaaS technology. As per automotive engineers autonomous vehicles are classified as below [21]:

1) Level 0 – hands and eyes of drivers used to run but no active help system
2) Level 1 – hands and eyes of drivers used to run, longitudinal or diagonal guide
3) Level 2 – hands and eyes were temporarily off and have road traffic control
4) Level 3 - hands and eyes were off and vehicles move based on deriver request
5) Level 4 – hands and mind were off condition and vehicles moved with no driver interference
6) Level 5 – driver not needed to drive the vehicle

Autonomous vehicles are operated based on a graphics processing unit (GPU) and sensors to control the vehicle. These vehicles used artificial intelligence which includes deep learning software. Autonomous vehicles were tested firstly in Germany and Japan with technical and traffic regulations. Several states in the US conduct road testing for autonomous vehicles. China framed the guidelines for road testing in 2017 and issued the license in 2018. Due to the technology enhancement, autonomous vehicles will be more in the future.

8.6 Smart Charging Influence on the Universal Energy System

From the above discussion, smart charging will be the key for electric vehicle charging, to maximize the possibility of synergies. In this section, the impact is analyzed based on system-wide impact and local system grid impact.

8.6.1 System-wide impact

Renewable energy integration with the power grid is an important concern for smart charging. Four important scenarios were enabled as a key for the integration as discussed below:

1) Business as usual scenario (BAU): It assumes that the advancement in the power system will be limited with the assumptions of EV deployment. Therefore, the charging patterns of EVs are unrestrained.

2) Partial smart charging scenario (V1G): It undertakes that the EV integration with the grid only in unidirectional smart charging.
3) Fully smart charging scenario (V2G): It undertakes that the EV connected with the grid in a bidirectional manner, that is, grid connected with the vehicle in one direction for charging and vehicle connected with the grid as a source while in parking or night-time.
4) MaaS smart charging scenario: It assumes a fully developed power system by the innovation, and this will develop toward mobility as a service (MaaS) for the evolution of EV technology toward autonomous vehicles.

More renewable energy generation (solar and wind) helps in reduction of electricity cost and peak load.

8.6.2 Local distribution grid impact

More adoption of EVs creates electricity problems with connections to the grid. Bottlenecks may occur while transmission, distribution, and transformers are not able to manage all the load requirements. This will create local distribution problems for the home appliances. To avoid such problems:

1) every home charging point consumption should report to the local distributors;
2) measures the load at the point of time;
3) communication equipment used to reduce the load, for example, the required load 16 A to 8 A for high charging time.

Slow charging will save the distribution network and cost departure. Smart charging allows substantial savings compared to business models.

8.7 Conclusion

This chapter work focuses on smart charging methods for an electric vehicle. The development of charging technologies in different countries was discussed. Every country formed different policies for the benefit of the customer. To avoid high costs during bill payment, dealers formed these policies and they follow them effectively. In this way, smart charging will be a good future technology. Business outlooks explained very clearly further to support smart charging infrastructure developments and to involve more actors to support. The integration of renewable energy sources with the power system has been highly supported for charging. Software used for the calculation of energy consumption by MaaS is explained clearly. To conclude, smart charging technology should satisfy the customers and increase the number of electric vehicles in the future.

References

[1] International Energy Agency*(IEA), Clean Energy Ministerial, and Electric Vehicles Initiative (EVI) (June 2020)*. "Global EV Outlook 2020: Entering the decade of electric drive?". *IEA Publications*. Retrieved 15 June 2020. See Statistical Annex, pp. 247–252

[2] Loveday, Steven (17 January 2020). "FINAL UPDATE: Quarterly Plug-In EV Sales Scorecard". InsideEVs.com.

[3] Kane, Mark (24 January 2019). "US Plug-In Electric Car Sales Charted: December 2018". InsideEVs.com.

[4] "Veloz: 2018 banner year for EV sales in California; up 84% from 2017; 312% increase from January to December". VELOZ. Green Car Congress. 14 January 2019.

[5] California New Car Dealers Association (CNCDA) (February 2019). "New Vehicle Registrations Remained Above 2 million Units in 2018" (PDF). CNCDA.

[6] California New Car Dealers Association (CNCDA) (February 2020). "California Auto Outlook Covering Fourth Quarter 2019: State New Vehicle Market Predicted to Remain Strong in 2020" (PDF). CNCDA

[7] Norsk Elbilforening (Norwegian Electric Vehicle Association) (January 2020). "Norwegian EV market". Norsk Elbilforening

[8] Cobb, Jeff (17 January 2017). "Top 10 Plug-in Vehicle Adopting Countries of 2016". HybridCars.com.

[9] Norwegian Road Federation (OFV) (2 January 2020). "Bilsalget i 2019"[Car sales in 2019] (in Norwegian). OFV.

[10] Kane, Mark (7 October 2018). "10% Of Norway's Passenger Vehicles Are Plug Ins". InsideEVs.com.

[11] Zhang, X., Xie, J., Rao, R., & Liang, Y. (2014). Policy incentives for the adoption of electric vehicles across countries. Sustainability, 6(11), 8056-8078.

[12] T. Shiramagond and W. Lee, "Integration of Renewable Energy into Electric Vehicle Charging Infrastructure," 2018 IEEE International Smart Cities Conference (ISC2), Kansas City, MO, USA, 2018, pp. 1-7, doi:10.1109/ISC2.2018.8656981

[13] S. Deb, K. Kalita and P. Mahanta, "Review of impact of electric vehicle charging station on the power grid," 2017 International Conference on Technological Advancements in Power and Energy (TAP Energy), Kollam, 2017, pp. 1-6, doi:10.1109/TAPENERGY.2017.8397215.

[14] Shen, Y., Fang, W., Ye, F., & Kadoch, M. (2020). EV charging behavior analysis using hybrid intelligence for 5G smart grid. Electronics, 9(1), 80.

[15] Høj, J. C. M. L., Juhl, L. T., & Lindegaard, S. B. (2018). V2G—An Economic Gamechanger in E-Mobility?. World Electric Vehicle Journal, 9(3), 35.

[16] Dethlefs, T., Brunner, C., Preisler, T., Renke, O., Renz, W., & Schröder, A. (2015, November). Energy service description for capabilities of distributed energy resources. In DA-CH Conference on Energy Informatics (pp. 24-35). Springer, Cham.

[17] Koch-Ciobotaru, C., Saez-de-Ibarra, A., Martinez-Laserna, E., Stroe, D. I., Swierczynski, M., & Rodriguez, P. (2015, September). Second life battery energy storage system for enhancing renewable energy grid integration. In 2015 IEEE Energy Conversion Congress and Exposition (ECCE) (pp. 78-84). IEEE.

[18] Chen, B., Hardy, K. S., Harper, J. D., Bohn, T. P., & Dobrzynski, D. S. (2015, June). Towards standardized vehicle grid integration: Current status, challenges, and next steps. In 2015 IEEE Transportation Electrification Conference and Expo (ITEC) (pp. 1-6). IEEE.

[19] Diao, Q., Sun, W., Yuan, X., Li, L., & Zheng, Z. (2016). Life-cycle private-cost-based competitiveness analysis of electric vehicles in China considering the intangible cost of traffic policies. Applied Energy, 178, 567-578.

[20] Li, Y., & Voege, T. (2017). Mobility as a service (MaaS): Challenges of implementation and policy required. Journal of transportation technologies, 7(2), 95-106.

[21] Leira, F. S., Johansen, T. A., & Fossen, T. I. (2015, March). Automatic detection, classification and tracking of objects in the ocean surface from UAVs using a thermal camera. In 2015 IEEE aerospace conference (pp. 1-10). IEEE.

Index

A

active current 47, 48, 89, 94, 108, 139
active filtration 91, 98, 111, 112, 120, 140
advanced metering 189, 190, 191, 196, 197, 215
apparent power 89, 93, 97, 99, 102, 111, 114, 131, 146
autonomous vehicles 226, 235, 236

B

batteries 2, 16, 17, 18, 19, 58, 213, 220, 231, 233
battery management systems (BMSs) 2
battery swap station 43

D

DC-DC converters 27, 30, 53, 170, 209
distributed energy resources 147, 149, 156, 164, 189
distributed energy storage 147, 155, 156
dynamic wireless charging 70, 74, 79, 80, 81, 82, 85

E

efficiency analysis 183
electric vehicle (EV) 39, 41, 69, 213, 220
e-mobility 219, 220, 222, 223, 230, 231, 232, 234
energy management systems 40, 189, 190, 201, 203, 206
energy storage systems (ESS) 1, 3, 7, 155, 208
EV battery charger 39, 40, 54

H

hierarchical control 92, 157, 158, 159, 164
hybrid electric vehicles 1, 40, 232

I

inductive charging 39, 45, 69, 74, 75, 76, 78, 84
internet-of-things 189

L

Lithium-ion battery (LIB) 3, 12, 13, 25, 31, 57
loads 28, 59, 91, 123, 147, 154, 155, 215

M

magnetic resonance 69, 71, 72, 74, 85
maximum power point tracking 185, 186, 187
microgrids 91, 147, 149, 152, 154, 189
multilevel inverter 169, 172, 173, 184

O

orthogonal components 90, 110, 123, 124, 126, 140

P

policies 149, 152, 219, 222, 236

power converters 56, 62, 189, 209, 210, 211

power electronic converters 26, 154, 189, 190, 207, 209, 210

PV panels 172

R

renewable energy sources 43, 59, 62, 147, 150, 155, 210, 211, 220, 236

S

smart battery charger 40, 58

smart grids 189, 190, 207, 215, 216

smart meters 190, 191, 193, 194, 196, 200, 204, 215

T

thermal management strategies 57

V

vehicle to grid technology 213

W

wireless charging 39, 69, 71, 74, 76, 78, 79, 80, 81, 210, 229

wireless charging of electric vehicles 87

wireless power transfer 44, 72, 74, 86

About the Editors

Nandhini Gayathri M. received her Bachelor's degree in Electrical and Electronics Engineering from SASTRA Deemed University, Thanjavur, India in 2005, her Master's degree in Power Electronics and Drives from SASTRA Deemed University, Thanjavur, India in 2007, and her Ph.D degree in Drives and Control from SASTRA Deemed University, Thanjavur, India in 2014. She has been working since 2007 as a Senior Assistant Professor with SASTRA Deemed University. She was funded by MHRD, India for the development of Pedagogy material for the course Digital Control of Electric Drives. She has authored over 50 scientific papers. She is a member of Institution of Engineering and Technology (IET), Indian Society for Technical Education (ISTE) and Indian Society of Systems for Science & Engineering (ISSE). Her main research interests include power converters, electric vehicle, wireless charging, control and design of electric machines.

Sanjeevikumar Padmanaban received his Ph.D. degree in electrical engineering from the University of Bologna, Bologna, Italy in 2012. He was an associate professor at VIT University from 2012 to 2013. In 2013, he joined the National Institute of Technology, India, as a faculty member. In 2014, he was invited as a visiting researcher at the Department of Electrical Engineering, Qatar University, Doha, Qatar, funded by the Qatar National Research Foundation (Government of Qatar). He continued his research activities with the Dublin Institute of Technology, Dublin, Ireland, in 2014. Furthermore, he served as an associate professor with the Department of Electrical and Electronics Engineering, University of Johannesburg, Johannesburg, South Africa, from 2016 to 2018. From March 2018 to February 2021, he has been a faculty member with the department of energy technology, Aalborg University, Esbjerg, Denmark. Since March 2021, he has worked as a professor with the CTIF Global Capsule (CGC) Laboratory, Department of Business Development and Technology, Aarhus University, Herning, Denmark. S. Padmanaban has authored over 300 scientific papers and was the recipient of the Best Paper-cum-Most Excellence Research Paper Award from

IET-SEISCON'13, IET-CEAT'16, IEEE-EECSI'19, IEEECENCON'19, and five best paper awards from ETAEERE'16 sponsored Lecture Notes in Electrical Engineering, Springer book. He is a fellow of the Institution of Engineers, India, the Institution of Electronics and Telecommunication Engineers, India, and the Institution of Engineering and Technology, UK. He is an editor/associate editor/editorial board for refereed journals, in particular, the IEEE SYSTEMS JOURNAL, IEEE Transactions on Industry Applications, IEEE ACCESS, IET Power Electronics, IET Electronics Letters, and Wiley-International Transactions on Electrical Energy Systems. He is the subject Editorial Board Member, Energy Sources, Energies Journal, MDPI, and the subject editor for the IET Renewable Power Generation, IET Generation, Transmission and Distribution and FACTS journal (Canada).

For Product Safety Concerns and Information please contact our
EU representative GPSR@taylorandfrancis.com Taylor & Francis
Verlag GmbH, Kaufingerstraße 24, 80331 München, Germany